VOLUME 3
RANDOM COUNTS IN SCIENTIFIC WORK

THE PENN STATE STATISTICS SERIES

An International Series in Statistics and Applications

General Editor: G. P. Patil, Professor of Statistics
The Pennsylvania State University

This new series brings volumes of importance to statisticians and scientists in related disciplines.
The series is organized in three general categories:

I. Monographs on Statistical Distributions and Applications

II. Monographs on Related Topics in Statistics

III. Edited multi-author volumes on subjects of contemporary interest in statistics

RANDOM COUNTS IN SCIENTIFIC WORK

Expanded from the Proceedings of the
Biometric Society Symposium
Dallas, Texas, December, 1968
(American Association for the
Advancement of Science Conference)

VOLUME 3
RANDOM COUNTS IN PHYSICAL SCIENCES, GEOSCIENCE, AND BUSINESS

G. P. PATIL, Editor
Department of Statistics
The Pennsylvania State University

The Pennsylvania State University Press
University Park and London

Library of Congress Catalogue Card Number 73-114351

Standard Book Number 271-00116-X

Printed in the United States of America

The Pennsylvania State University Press
University Park, Pennsylvania 16802

The Pennsylvania State University Press, Ltd.
London W. 1

CONTENTS

PREFACE

As Professor C. R. Rao has put it: "Endowed with a wide and rapidly expanding literature, both in diverse fields of applied science and statistics, the subject of random counts in scientific work or discrete distributions and their applications offers a fascinating area of study and research." It is of course conceivable that older and more familiar theories of statistical inference may be specifically needed or used to analyze exclusively the data on random counts. However, it is evident that new and novel mathematical and statistical approaches and procedures pertaining to special types of counted data related to individual scientific problems are long overdue.

These new procedures may be evolved in two ways. One method is to investigate in detail individual discrete distributions for their mathematical and statistical properties and for their natural applications. The second approach is to concentrate on a scientific problem that generates or requires counted data and then create and coordinate appropriate discrete models and methods so as to arrive at a meaningful solution of the scientific problem.

Random counts thus has much to offer for the advancement of the basic and applied sciences, as well as for the advancement of statistics itself.

From this premise and the multidisciplinary background of the annual conferences of the American Association for the Advancement of Science, I had felt for some time that the Biometric Society and the Statistics Section of the AAAS would provide the proper atmosphere for a much-needed program on the subject of random counts in scientific work.

During the time that I served on the program committee of the Biometric Society and while I was the program chairman for 1968 meetings, I had several instructive and encouraging discussions with the immediate past program chairmen D. S. Robson and E. C. Pielou, with H. O. Hartley, Biometric Society ENAR President, and with C. I. Bliss, the Vice-President of the Statistics Section of the AAAS. A symposium on the subject of random counts was planned and organized at the Dallas Conference of the Biometric Society in December of 1968; twenty participants were on the program.

The symposium was a great success in that there was a very

lively exchange and interaction between scientists from different fields and statisticians interested in the analysis of random counts. Although there seemed to be divergence of levels and diversity in the presentations, this turned out to be more of an asset than a liability.

Because of the quality and value of the presented papers, as well as widespread interest in the topic, the participants and audience felt that the proceedings should be published. It also seemed appropriate to invite additional papers from specialists who were not present in Dallas. Twenty-two papers were invited and organized in this category.

For purposes of convenience, the final total of 42 papers has been divided into three discrete subject-matter groups, published in a like number of volumes under the general title *Random Counts in Scientific Work*, treating random counts in (1) models and structures, (2) biomedical and social sciences, and (3) physical sciences, geoscience, and business. The present volume, *Random Counts in Physical Sciences, Geoscience, and Business*, represents the third of the three volumes published. It is expected that more volumes will be added to this series as progress and new results in the field of random counts in scientific work come about. Any comments and suggestions from readers would be most welcome.

University Park, Pennsylvania G. P. Patil
September 1970

DISCRETE DISTRIBUTION MODELS ARISING IN POLLEN STUDIES

JAMES E. MOSIMANN
National Institutes of Health

SUMMARY

Counts of the numbers of various kinds of pollen found at different depths in sediment are used in paleoecological studies of the earth's history. This paper presents various models for use in such studies.

First, models for variation in counts at a given depth are presented. In these the variation in counts depends on the type of sampling used by the investigator, and the familiar multinomial, negative multinomial and Poisson distributions arise naturally.

Next, variation from depth to depth is discussed. This variation is induced by the history of the vegetation along with natural events which affect the dispersal and preservation of pollen. Two types of models, (I) Fixed Effects and (II) Random Effects are discussed. In particular, a set of null models is presented. These include: Null Model A, the case where the population pollen proportions are constant from depth to depth; Null Model B, where the population pollen concentrations vary as independent gamma variables (implying that the proportions vary as in the Dirichlet or multivariate Beta distribution); and Null Model C, the case where the pollen concentrations vary as independent Poisson variables. In Null Model B the Dirichlet-compound multinomial and Dirichlet-compound negative multinomial models arise naturally.

The use of a multivariate lognormal model is briefly examined.

1. INTRODUCTION

Counts of the numbers of the various kinds of fossil pollen grains found at different depths in accumulated sediment form a fossil pollen profile whose characteristics reflect the past changes in vegetation and climate which occurred in the area of sediment deposition. At a given depth in the sediment, the numbers of the various kinds of preserved pollen reflect in a rough way the numbers and kinds of wind-pollinated plants which grew in the area when the sediment at the particular depth was deposited (Fig. 1). Thus by considering changes in the numbers of pollen with changes in depth one can attempt a reconstruction of past changes in vegetation through time.

Interesting recent examples of the use of pollen to interpret the past can be found in Tauber [17-19] who has studied aerial dispersion models and data; in Davis [2-4] who has succeeded in measuring pollen deposition rates (Number of grains per square centimeter per year) for various pollen types in postglacial sediments in Connecticut; in Wright [22] who has provided a careful reconstruction of the history of pine in postglacial time in Minnesota; in Livingstone [6] who has used correction factors to allow for the disproportionate representation of pollen of various kinds of plants due to their different pollen production and dispersal capabilities in the postglacial history of Eastern Canada; and in Whitehead and Tan [21] who have compared fossil pollen profiles with modern pollen percentages in North Carolina. A wealth of morphological diversity is revealed in scanning electron microscope photographs of pollen grains of various plants, Martin and Drew [7].

Various discrete probability models arise naturally in pollen work. In estimating the actual number of pollen grains in a sample by aliquots, the Poisson model has been used (Davis [2]) with the same justification as is common in bacteriological studies (Pettigrew and Mohler [16]). In interpreting pollen profiles themselves, multinomial and negative multinomial as well as compound multinomial and compound negative multinomial distributions arise naturally (Mosimann [10, 11]). The purposes of the present paper are (1) to indicate briefly how these distributions arise in pollen work; (2) to indicate various models for interpreting pollen profiles, with in particular a consideration of the compounding model (Model II(a) in Section 3) in pollen applications; (3) to consider null models and covariation properties of counts given the null model; and (4) to briefly consider some properties of the lognormal distribution which may render it a useful model in analyzing pollen counts.

In summary, the aim of this paper is to bring together in one place and in a consistent notation a number of models for pollen study. The most important model development, involving time or depth-dependent models (type II(b) in Section 3), remains to be done. The models given here present a beginning point for what promises to be a most interesting area of applications of discrete distributions in biology and geology.

2. MULTINOMIAL AND NEGATIVE MULTINOMIAL DISTRIBUTIONS

Notation for the major types of variables discussed in this paper is given in Figure 2. The concentration per unit volume of sediment of each of k types of pollen in the population of preserved pollen at the ith depth or level of a sedimentary profile will be denoted by

$$\underset{\sim}{x}_i = (x_{i1}, x_{i2}, \ldots, x_{ik}),$$

with the corresponding population proportions at the ith level being denoted

$$\underset{\sim}{p}_i = (p_{i1}, p_{i2}, \ldots, p_{ik}),$$

where $p_{ij} = x_{ij} / \sum_{j=1}^{k} x_{ij}$; $j = 1,\ldots,k$.

The corresponding population ratios are denoted

$$\underset{\sim}{u}_i = (u_{i1}, u_{i2}, \ldots, u_{ik}),$$

where $u_{ij} = x_{ij}/x_{ik} = p_{ij}/p_{ik}$; $j = 1, \ldots, k$.

The x's, p's, and u's describe the population of preserved pollen at ith level. Sample counts of pollen from the ith level will be denoted by

$$y_{i1}, y_{i2}, \ldots, y_{ik} .$$

The y's will always be considered as values of random variables. The x's, p's and u's will be considered as parameters in some models and as values of random variables in others.

Sampling Rules.

When obtaining sample counts from a given (ith) level three stopping rules are used with some frequency. In the first, more common case, counts of each of various kinds of grains are continued until a fixed total number, n, of grains is reached. Given generalized Bernoulli Trials, counts at the ith level would follow a multinomial distribution with

$$m(y_{i1}, \ldots, y_{ik}) = \binom{n}{y_{i1}, y_{i2}, \ldots, y_{ik}} \prod_{j=1}^{k} p_{ij}^{y_{ij}} , \qquad (1)$$

where $\sum_{j=1}^{k} y_{ij} = n$ and $p_{ij} = x_{ij} / \sum_{j=1}^{k} x_{ij}$, $i = 1, \ldots, k$.

The second stopping rule is to count the various kinds of grains until a fixed number, r, of the kth pollen type is obtained. Given generalized Bernoulli Trials, counts at the ith level would follow a negative multinomial distribution with

$$nm(y_{i1}, \ldots, y_{ik}) = \binom{r-1 + \sum_{j=1}^{k-1} y_{ij}}{y_{i1}, y_{i2}, \ldots, y_{ik-1}, r-1} \times$$

$$p_k^{\,r} \prod_{j=1}^{k-1} p_{ij}^{y_{ij}} , \qquad (2)$$

where the p's are defined as above.
Negative multinomial counts range over 0,1,2,..., and negative multinomial sampling will be called "inverse sampling" in this paper. Techniques for analyzing either multinomial or negative multinomial pollen counts are given in Mosimann [12].

A third stopping rule, used in the estimation of the x's, the pollen concentrations, is to count all the grains in a given aliquot representing a fraction of a suspension of pollen from sediment of known volume at the ith level. As noted by Davis [2] the number found of the jth type of grain would follow a Poisson distribution with mean x_{ij}. With independent mixing of the grains of different types in the original suspension, the joint distribution of the counts $y_{i1} \ldots y_{ik}$ would be the product of k independent Poisson distributions with

$$p(y_{i1}, \ldots, y_{ik}) = \prod_{j=1}^{k} \frac{e^{-x_{ij}} x_{ij}^{y_{ij}}}{y_{ij}!} \quad . \tag{3}$$

Counts obtained in this manner will be called "independent free counts" in this paper. The important contribution of Davis consists not only in obtaining such counts, but in controlled dating of the sediment so that counts per unit of sediment can be converted to counts per surface area per time.

Moments at the ith Level.

For the ith level, moments of fixed - n multinomial counts are readily and habitually expressed in terms of p's:

$$E(y_{ij} | \underset{\sim}{x}_i) = n x_{ij} / \sum_j x_{ij} = n p_{ij}, \quad j = 1, \ldots, k \tag{4}$$

$$Var(y_{ij} | \underset{\sim}{x}_i) = n p_{ij}(1 - p_{ij}) \quad , \quad j = 1, \ldots, k \tag{5}$$

$$Cov(y_{ij}, y_{im} | \underset{\sim}{x}_i) = -n p_{ij} p_{im} \quad , \quad j \neq m \ . \tag{6}$$

The p's are the proportions of pollen which are preserved at the ith level. Negative multinomial moments for the inverse counts at the ith level are readily expressed in terms of u's:

$$E(y_{ij} | \underset{\sim}{x}_i) = r x_{ij} / x_{ik} = r p_{ij} / p_{ik} = r u_{ij} \qquad j = 1, \ldots, k-1 \tag{7}$$

$$Var(y_{ij} | \underset{\sim}{x}_i) = r u_{ij}(1 + u_{ij}) \qquad j = 1, \ldots, k-1 \tag{8}$$

$$Cov(y_{ij}, y_{im} | \underset{\sim}{x}_i) = + r u_{ij} u_{im} \qquad j \neq m \ . \tag{9}$$

Negative multinomial moments are easily remembered from multinomial moments by substituting u's for p's, r for n, and plus signs for minus signs.

In both the multinomial and negative multinomial the parameters for a given level can be expressed in terms of x's, the numbers of preserved pollen grains. For years it has been convenient to express multinomial moments in terms of probabilities or proportions. In an analogous way, u's, or odds ratios, are convenient parameters in the negative multinomial case. A negative multinomial count, y_{ij}, gives the number of type j obtained while counting to r grains of type k. When y_{ij} is divided by r, a sample odds ratio of type j to type k is obtained.

Moments for independent free counts from the ith level are

those of independent Poisson variables:

$$E(y_{ij}|\underset{\sim}{x}_i) = x_{ij} , \qquad j = 1, \ldots, k \tag{10}$$

$$Var(y_{ij}|\underset{\sim}{x}_i) = x_{ij} , \qquad j = 1, \ldots, k \tag{11}$$

$$Cov(y_{ij}, y_{im}|\underset{\sim}{x}_i) = 0, \quad j \neq m . \tag{12}$$

While variation at a given (ith) depth is important, the paleoecological problem generally is one of determining changes in x's (or p's or u's) from depth to depth in a profile. The next set of models considered will allow for depth-to-depth variation in preserved pollen.

3. VARIATION WITH DEPTH

Depending on the real vegetation history and also on the spacing of samples by depth, one can consider various models for depth-to-depth variation in a single profile. There is no single desired approach in practice. A useful approach is to consider models in a fashion analogous to the standard analysis of variance, without, however, any assumption of normal distributions.

I. The Fixed-Effects Model

In this model the $\underset{\sim}{x}_i$'s at the various levels are regarded as fixed vectors, and are not considered as values of random variables; instead, the $\underset{\sim}{x}_i$'s are taken as parameters which change with depth. Various statistics computed from sample counts ($\underset{\sim}{y}_i$'s) across depths have descriptive value for the single profile under consideration.

II. The Random Effects Model

In this model the $\underset{\sim}{x}_i$'s themselves are considered as values of random variables with some probability distribution according to which values of $\underset{\sim}{x}_i$'s are determined. Sample counts can then be used to estimate parameters of the probability distribution describing between-depth variation. Models of this type can be subdivided:

II(a) Independently determined $\underset{\sim}{x}$'s, $\underset{\sim}{p}$'s or $\underset{\sim}{u}$'s from level to level.

In this case either the spacing of samples by depth or the vegetation history itself is such that there are independent determinations of the $\underset{\sim}{x}_i$ from depth to depth.

II(b) Non-independently determined $\underset{\sim}{x}$'s, $\underset{\sim}{p}$'s or $\underset{\sim}{u}$'s from level to level

In this case the spacing or samples by depth or the actual vegetation history, is such that the determination of $\underset{\sim}{x}_i$'s (or $\underset{\sim}{p}_i$'s or $\underset{\sim}{u}_i$'s) at one level is not independent from sample to sample with depth. Perhaps there is some autocorrelation function linking the value of $\underset{\sim}{x}_{i+1}$ to $\underset{\sim}{x}_i$ and to other $\underset{\sim}{x}$'s in previous time levels.

Various applications of these models are illustrated in Figs. 3-6. In Model II(a) there is no time trend effect, and the models in this group seem most useful as null models, rather than as models to describe temporal variation. However, for practical purposes, independent levels might be obtained with incomplete sampling in a long core and Model II(a) may apply descriptively in such a case.

Many postglacial cores in eastern North America cover a relatively short time period (say, less than 14,000 years) and time trends are clearly in evidence. The descriptive approach of Model I or the approach of Model II(b) seem called for. Thus far, the use of Model I seems adequate for many purposes in such cores.

Moments under Model II(a).

For the remainder of this paper we shall deal with Model II(a), the random effects independent levels model. Moment results for this model are given below (analogous formulas also can be obtained under the fixed effects Model I). Model II(a) gives us familiar compound distributions, the compound multinomial and compound negative multinomial.

With fixed-n multinomial counts at the ith level, then across all levels we would have the following moments:

$$E(Y_i) = nE(X_j/\Sigma X) = nE(P_j), \qquad j = 1, \ldots, k \quad (13)$$

$$Var(Y_j) = nE(P_j)[1-E(P_j)] + n(n-1)Var(P_j) \qquad j = 1, \ldots, k \quad (14)$$

$$Cov(Y_j,Y_m) = -nE(P_j)E(P_m) + n(n-1)Cov(P_j,P_m), \qquad j \neq m \, . \quad (15)$$

Since the P's are bounded random variables, all above moments of the P's are finite.

With negative multinomial inverse counts at the ith level, we would have across levels

$$E(Y_j) = rE(X_j/X_k) = rE(U_j), \qquad j = 1, \ldots, k-1 \quad (16)$$

$$Var(Y_j) = rE(U_j)[1+E(U_j)] + r(r+1)Var(U_j), \qquad j = 1, \ldots, k-1 \quad (17)$$

$$Cov(Y_j,Y_m) = rE(U_j)E(U_m) + r(r+1)Cov(U_j,U_m), \; j \neq m , \quad (18)$$

where $E(U_j)$, $Var(U_j)$, and $Cov(U_j,U_m)$ which are not necessarily finite are assumed to be so in the above expressions.

Again, where one is estimating pollen concentration with independent free Poisson counts at the ith level, across level moments are

$$E(Y_j) = E(X_j), \qquad j = 1, \ldots, k \quad (19)$$

$$Var(Y_j) = E(X_j) + Var(X_j), \qquad j = 1, \ldots, k \quad (20)$$

$$Cov(Y_j,Y_m) = Cov(X_j,X_m), \qquad j \neq m . \quad (21)$$

where, as before, the relevant moments of the x's are assumed finite.

4. NULL MODELS

In interpreting a pollen profile the joint variation of the counts of the various pollen types is usually evaluated visually from a diagram. In practice, because many important events are revealed as short term time trends, models of type II(b) or type I may be more applicable than those of type II(a). However, a particularly useful kind of type II(a) model is the null model, which describes how pollen proportions would vary "by chance alone", that is, in essentially random situations not reflecting any vegetation trends nor orderly vegetation history. The need for consideration of such null models stems from the fact that covariation of counts from the essentially random situations is complex and is easily subject to misinterpretation as reflecting special relations among taxa and special vegetation history. Two null models have been described elsewhere (Mosimann [11], Martin and Mosimann [8]). These will be reconsidered here with the additional consideration of variation of independent free counts as obtained by Davis [2]. Other possible null models will also be mentioned.

Understanding the variation expected in these null situations permits a clarity of thought in examining data which vary in a complex non-null manner.

Null Model A.

This model describes the random behavior of counts where the proportions of pollen do not vary from depth to depth. In this model the concentration vector X may itself be taken as constant across levels or may be any random vector variable such that

$$X_j = a_j X_k \qquad j = 1, \ldots, k, \tag{22}$$

where we set $a_k = 1$. Thus $\tilde{P}$ and $\tilde{U}$ are degenerate random vectors which with probability 1 take only the values

$$\tilde{p} = (a_1/\Sigma a, \ldots, a_k/\Sigma a)$$

and

$$\tilde{u} = (a_1, \ldots, a_k),$$

respectively. The situation described is the only one in which there is no change in the population proportions or ratios with depth. The situation described includes, of course, the null case of constant concentrations, x's.

Where we assume $Var(X_k)$ to be finite we can write the across-level moments of the X's as

$$E(X_j) = a_j E(X_k), \qquad j = 1, \ldots, k \tag{23}$$

$$Var(X_j) = a_j^2 Var(X_k), \qquad j = 1, \ldots, k \tag{24}$$

$$Cov(X_j, X_m) = a_j a_m Var(X_k), \qquad j \neq m . \tag{25}$$

Of course X_j and X_m are perfectly correlated in this null situation.

Sample Counts Under Null Model A.

Under Null Model A, fixed-n counts follow a multinomial distribution across levels and inverse counts follow a negative multinomial distribution across levels. Moments of these distributions may be obtained from equations (4), (5) and (6) by substituting $a_j/\Sigma a$ for $x_{ij}/\Sigma x$ and from equations (7), (8), and (9) by substituting a_j for x_{ij}/x_{ik}. The distribution of independent free counts depends on the distribution of X_k. Moments across levels are

$$E(Y_j) = a_j E(X_k), \qquad j = 1, \ldots, k \tag{26}$$

$$Var(Y_j) = a_j E(X_k) + a_j^2 Var(X_k), \qquad j = 1, \ldots, k \tag{27}$$

$$\mathrm{Cov}(Y_j, Y_m) = a_j a_m \mathrm{Var} X_k, \qquad j \neq m . \tag{28}$$

It is possible to obtain exact results for the correlation of proportions formed from independent free counts under Null Model A. The results hold for exactly those conditions in practice for which proportions can be calculated; that is, they are conditional upon $\Sigma Y \neq 0$.

Under Null Model A the joint probability mass function of the counts, given the value of X_k, can be written

$$p(y_1, \ldots, y_k | x_k) = \prod_{j=1}^{k} \frac{(a_j x_k)^{y_j} e^{-a_j x_k}}{y_j !} ,$$

and the conditional distribution given Σy is, as is well-known, the multinomial with

$$m(y_1, \ldots, y_k | x_k, \Sigma y) = \binom{\Sigma y}{y_1, \ldots, y_k} \prod_{j=1}^{k} p_j^{y_j} ,$$

where we write $P_j = a_j / \Sigma a$. Note that the x_k cancels in the ratio $a_j x_k / (x_k \Sigma a)$. Thus one has the conditional expectations

$$\begin{aligned} E(Y_j/\Sigma Y | x_k, \Sigma y) &= (1/\Sigma y) E(Y_j | x_k, \Sigma y) \\ &= p_j . \end{aligned} \tag{29}$$

Similarly

$$\begin{aligned} E[(Y_j/\Sigma Y)^2 | x_k, \Sigma y] &= (1/\Sigma y)^2 [(\Sigma y) p_j (1-p_j) + (\Sigma y)^2 p_j^{\,2}] \\ &= (1/\Sigma y) p_j (1-p_j) + p_j^{\,2} ; \end{aligned} \tag{30}$$

and also

$$E([Y_i/\Sigma Y]\,[Y_j/\Sigma Y] | x_k, \Sigma y) = -(1/\Sigma Y)\, p_j p_m + P_j p_m . \tag{31}$$

The variable ΣY has a Poisson distribution with mean $x_k \Sigma a$. We can write $h(\Sigma y | x_k)$ for the probability mass function of ΣY given that $\Sigma Y \neq 0$. Thus $\sum_{\Sigma y=1}^{\infty} h(\Sigma y | x_k) = 1$. It is clear that $\sum_{\Sigma y=1}^{\infty} (1/\Sigma y)\, h\,(\Sigma y | x_k)$ is finite. This is some function of x_k, say, $w(x_k)$. Then, for example,

$$E[Y_j^{\,2}/(\Sigma Y)^2] = \underset{x_k}{E}\, [w(x_k)] p_j p_m + p_j p_m .$$

The constants p_j and p_m are clearly finite. Since $Y_j/\Sigma Y$ is a bounded random variable, $E[Y_j^2/(\Sigma Y)^2]$ must likewise be finite, and hence $\underset{x_k}{E}[w(x_k)]$ must also be finite. Without further ado, then, we can write

$$\mathrm{Var}(Y_j/\Sigma Y) = p_j(1-p_j)\underset{x_k}{E}[w(x_k)], \qquad j = 1, \ldots, k \tag{32}$$

$$\mathrm{Cov}[(Y_j/\Sigma Y), (Y_m/\Sigma Y)] = -p_j p_m \underset{x_k}{E}[w(x_k)], \quad j \neq m . \tag{33}$$

Thus the correlation coefficient for proportions obtained from independent Poisson counts under Null Model A is:

$$\mathrm{Cor}\left(\frac{Y_j}{\Sigma Y}, \frac{Y_m}{\Sigma Y}\right) = -\sqrt{\frac{p_j\, p_m}{(1-p_j)(1-p_m)}}\,, \tag{34}$$

which is the same as that for fixed-n multinomial counts (cf. Table 1). Correlation coefficients for the three kinds of stopping rules under Null Model A are given in Table 1.

Null Model B.

This model describes the behavior of sample counts where the proportions of preserved pollen vary from depth to depth but in an "essentially random" manner to be specified here.

In Null Model B the concentration vector $\underset{\sim}{X}$ may be any random vector such that (1) $X_1, \ldots, X_k$ are positive, nondegenerate, and mutually independent random variables and (2) the vector of their proportions, $\underset{\sim}{p}$, is independent of ΣX. But these two conditions hold if and only if each X_j, $j = 1, \ldots, k$, has a gamma distribution with the same scale parameter (Mosimann [10]). (The vector of ratios, $\underset{\sim}{U}$ is independent of ΣX if and only if $\underset{\sim}{p}$ is independent of ΣX.) Thus in Null Model B the joint density function of $X_1, \ldots, X_k$ is given by

$$f(x_1, \ldots, x_k) = \prod_{j=1}^{k} \frac{\alpha^{b_j}}{\Gamma(b_j)} x_j^{b_j - 1} e^{-\alpha x_j}, \tag{35}$$

where $\alpha, b_j > 0$, $j = 1, \ldots, k$.

Moments of the pollen concentrations under Null Model B are

$$E(X_j) = b_j/\alpha, \qquad j = 1, \ldots, k \tag{36}$$

$$\mathrm{Var}(X_j) = b_j/\alpha^2, \qquad j = 1, \ldots, k \tag{37}$$

$$\mathrm{Cov}(X_j, X_m) = 0, \qquad j \neq m . \tag{38}$$

Under Null Model B the proportions follow a singular Dirichlet or multivariate beta distribution with density function

$$g(p_1, \ldots, p_k) = [\Gamma(\Sigma b)/\Pi\Gamma(b_j)] \prod_{j=1}^{k} p_j^{b_j - 1} , \quad (39)$$

where the parameters are as in (35). The moments of the proportions under this model are

$$E(P_j) = b_j/\Sigma b = E(X_j)/E(\Sigma X), \quad j = 1, \ldots, k \quad (40)$$

$$Var(P_j) = E(P_j)\,[1-E(P_j)]\,[1/(1+\Sigma b)], \quad j = 1, \ldots, k \quad (41)$$

$$Cov(P_j, P_m) = -E(P_j)E(P_m)\,[1/(1+\Sigma b)], \quad j \neq m . \quad (42)$$

The ratios $U_1, \ldots, U_k$ under Null Model B follow a singular inverted Dirichlet distribution (Tiao and Cuttman, [20]) with density function

$$h(u_1, \ldots, u_{k-1}) = [\Gamma(\Sigma b)/\Pi\Gamma(b)] \prod_{j=1}^{k} [u_j^{b_j - 1}/(\Sigma u)^{b_j}] , \quad (43)$$

where $u_k \equiv 1$. Moments of ratios with $b_k > 2$ are

$$E(U_j) = b_j/(b_k-1), \quad j = 1, \ldots, k-1 \quad (44)$$

$$Var(U_j) = E(U_j)\,[1+E(U_j)]/(b_k-2) \quad j = 1, \ldots, k-1 \quad (45)$$

$$Cov(U_j, U_m) = E(U_j)E(U_m)/(b_k-2) \quad j \neq m . \quad (46)$$

When $b_k > 2$, variances and covariances are finite, but are not so if $b_k < 2$. If $b_k > 1$, expectations of the U's are finite.

Under Null Model B, the number of pollen preserved of any type is determined independently of the number of any other type. Further, the proportion of pollen preserved of any type is determined independently of the total pollen preserved for all types together. [1]

---

[1] It may be worthwhile to illustrate a case of a proportion being correlated with the total pollen deposition, Σx. Suppose there is no pine in the vegetation surrounding a lake. Through time the size of a lake may vary and correspondingly the distance from shore of a given point of deposition. When the lake is smaller, more local pollen from the shore may reach the given site of deposition (Σx large). Small amounts of pine pollen, blowing in independently from a long distance away will form only a small proportion of pine. When the lake is large, much less local pollen reaches the site (Σx small) and the pine pollen blowing in from more remote areas forms an appreciable proportion of the total. Here the proportion of pine would be negatively correlated with Σx.

Any correlations among proportions of pollen preserved in this null situation may be regarded as induced by the constraint that proportions must total 1. No particular meaning, other than the presence of a constraint, would be attached to the covariation of counts in this situation. Null Model B is one expression of the notion of essentially random pollen deposition.

Sample Counts Under Null Model B.

Fixed-n counts have a Dirichlet- or Beta-compound multinomial distribution, and inverse counts have a Dirichlet- or Beta-compound negative multinomial distribution. Moments for both cases are obtained in Mosimann [11]. They are readily found here using equations (13) through (18) in conjunction with equations (40) through (46). They are, respectively: Dirichlet-compound multinomial (fixed-n counts):

$$E(Y_j) = n(b_j/\Sigma b) = nE(P_j) = nE(X_j)/E(\Sigma X) \qquad j = 1, \ldots, k-1 \qquad (47)$$

$$\mathrm{Var}(Y_j) = nE(P_j)(1-E(P_j))[(n+\Sigma b)/(1+\Sigma b)] \qquad j = 1, \ldots, k-1 \qquad (48)$$

$$\mathrm{Cov}(Y_j, Y_m) = -nE(P_j)E(P_m)[(n+\Sigma b)/(1+\Sigma b)] \qquad j \neq m\ ; \qquad (49)$$

Dirichlet-compound negative multinomial (r grains of type k needed to stop):

$$E(Y_j) = r[b_j/(b_k-1)] = rE(U_j), \qquad j = 1, \ldots, k-1 \quad (50)$$

$$\mathrm{Var}(Y_j) = rE(U_j)[1+E(U_j)][(b_k+r-1)/(b_k-1)], \qquad j = 1, \ldots, k-1 \quad (51)$$

$$\mathrm{Cov}(Y_j, Y_m) = rE(U_j)E(U_m)[(b_k+r-1)/(b_k-1)], \qquad j \neq m\ .$$

Here $b_k > 2$ is assumed. If $b_k < 2$, then variances and covariances are not finite. If $b_k < 1$, then $E(Y_j)$ is not finite.

The distribution of each of the independent free counts is that of a Poisson compounded with a gamma and thus is negative binomial. Writing the density function of each gamma distribution as indicated by equation (35), the corresponding negative binomial probability mass function is

$$nb(y_j) = \frac{\Gamma(y_j+b_j)}{\Gamma(b_j)\Gamma(y_j+1)} \left(\frac{1}{\alpha+1}\right)^{y_j} \left(\frac{\alpha}{\alpha+1}\right)^{b_j}, \qquad (53)$$

again keeping the same notation as in equation (35).

Moments of the free counts are

$$E(Y_j) = b_j/\alpha = E(X_j), \qquad j = 1, \ldots, k \quad (54)$$

$$\mathrm{Var}(Y_j) = b_j(1/\alpha)[1+(1/\alpha)] = E(X_j)[b_j+E(X_j)]/b_j, \quad j = 1, \ldots, k \quad (55)$$

$$\mathrm{Cov}(Y_j, Y_m) = 0, \qquad j = 1, \ldots, k \quad (56)$$

These could be obtained readily from equations (7), (8), and (9) noting that $u_{ij} = 1/\alpha$, $r = b_j$, and recalling that b_j here need not be an integer. They could also be obtained readily from (19), (20) and (21) in conjunction with (36), (37) and (38).

As in the case of Null Model A, it is possible to obtain exact results for the correlation of proportions formed from free sample counts. Again the results hold given $\Sigma Y \neq 0$. Under Null Model B the joint probability mass function is that of k independent negative binomials (with the same parameter α):

$$\prod_{j=1}^{k} \frac{\Gamma(y_j+b_j)}{\Gamma(b_j)\Gamma(y_j+1)} \left(\frac{1}{\alpha+1}\right)^{y_j} \left(\frac{\alpha}{\alpha+1}\right)^{b_j}, \qquad (57)$$

The probability mass function of ΣY is then

$$\frac{\Gamma(\Sigma y+\Sigma b)}{\Gamma(1+\Sigma y)\Gamma(\Sigma b)} \left(\frac{1}{\alpha+1}\right)^{\Sigma y} \left(\frac{\alpha}{\alpha+1}\right)^{\Sigma b}, \qquad (58)$$

so that the conditional distribution of $Y_1, \ldots, Y_k$, given ΣY, is given by

$$cm(y_1, \ldots, y_k|\Sigma y) = \frac{\Gamma(1+\Sigma y)\Gamma(\Sigma b)}{\Gamma(\Sigma y+\Sigma b)} \prod_{j=1}^{k} \frac{\Gamma(y_j+b_j)}{\Gamma(b_j)\Gamma(y_j+1)}, \qquad (59)$$

which is the probability mass function of the Dirichlet compound multinomial distribution (see, for example, Mosimann [10], p. 66]). Using (47) and writing p_j for $E(P_j) = b_j/\Sigma b$ for convenience, we find

$$E(Y_j/\Sigma Y|\Sigma y) = (1/\Sigma y)\ (\Sigma y)\ p_j = p_j, \qquad j = 1, \ldots, k \quad (60)$$

Using (47) and (48) and (49) together and writing $t(\Sigma y)$ for $(\Sigma y+\Sigma b)/(1+\Sigma b)$ we obtain

$$E[(Y_j/\Sigma Y)^2|\Sigma y] = \frac{t(\Sigma y)}{\Sigma y} p_j(1-p_j) + p_j^2, \qquad j = 1, \ldots, k \quad (61)$$

$$E[(Y_j/\Sigma Y)(Y_m/\Sigma Y)|\Sigma y) = -\frac{t(\Sigma y)}{\Sigma y} p_j p_m + p_j p_m, \; j \neq m. \qquad (62)$$

Taking expectations with respect to ΣY we find moments for sample proportions formed from independent free counts under Null Model B

$$E(Y_j/\Sigma Y) = p_j, \qquad j = 1, \ldots, k \tag{63}$$

$$Var(Y_j/\Sigma Y) = p_j(1-p_j) \underset{\Sigma y}{E} [t(\Sigma y)/\Sigma y], \qquad j = 1, \ldots, k \tag{64}$$

$$Cov[(Y_j/\Sigma Y),(Y_m/\Sigma Y)] = - p_j p_m \underset{\Sigma y}{E} [t(\Sigma y)/\Sigma y], \; j \neq m . \tag{65}$$

It is not difficult to show that $\underset{\Sigma y}{E} [t(\Sigma y)/\Sigma y]$ must be finite as in the analogous case in Null Model A (31). The correlation coefficient of proportions from independent free counts under Null Model B is the same as that for Null Model A [equation (34)]; and the same as that for fixed-n counts under null models A or B (Table 1).

Null Model C.

A third null model can be briefly examined. In this model pollen fall is thought of as "essentially random" as follows: the pollen concentrations $X_1, \ldots, X_k$ are mutually independent random variables each having a Poisson distribution with mean λ_j, respectively.

Moments of the X's are thus

$$E(X_j) = \lambda_j, \qquad j = 1, \ldots, k \tag{66}$$

$$Var(X_j) = \lambda_j, \qquad j = 1, \ldots, k \tag{67}$$

$$Cov(X_j, X_m) = 0, \; j \neq m . \tag{68}$$

It is possible to obtain the moments of the preserved pollen proportions, P_j, using the same argument which gave the moments of proportions formed from independent free counts under Null Model A. The results hold under the conditions that proportions are meaningful; namely given $\Sigma X \neq 0$. The mathematical argument follows immediately letting X_k in Null Model A be degenerate $P[X_k = c] = 1$. Thus under Null Model C

$$E(P_j) = \lambda_j/\Sigma\lambda = E(X_j)/E(\Sigma X), \qquad j = 1, \ldots, k \tag{69}$$

$$Var(P_j) = E(P_j)[1-E(P_j)]E(1/\Sigma X), \qquad j = 1, \ldots, k \tag{70}$$

$$Cov(P_j, P_m) = - E(P_j)E(P_m)E(1/\Sigma X), \qquad j \neq m . \tag{71}$$

The correlation of preserved pollen proportions is as in (34) [denoting $E(P_j)$ by p_j) and is the same as that for the P's under Null Model B (Table 1). The conditional argument yielding equation (34) holds for proportions formed from any distribution of the discrete random variables $X_1, \ldots, X_k$, whose conditional distribution, given ΣX(and given $\Sigma X \neq 0$), is multinomial with p's constant with respect to any further expectations.

Among multivariate generalized power series distributions (Patil [14]), Joshi and Patil [5] have shown that the distribution of $X_1, \ldots, X_k$ is sum-symmetric (Patil [15]) if and only if the conditional distribution, given Σx, is multinomial. Subject to the restriction on the multinomial p's just noted, then, a sum-symmetric generalized power series distribution will yield proportions correlated as in (34). (From the section on independent free counts for Null Model B, we see that proportions can be correlated as in (34), where the distribution is not sum-symmetric.)

Sample Counts Under Null Model C.

We consider briefly the case of fixed-n counts. Using (13), (14) and (15) in conjunction with (69), (70) and (71) we have (denoting $E(P_j) = p_j$)

$$E(Y_j) = nE(P_j) = p_j, \qquad j = 1, \ldots, k \tag{72}$$

$$Var(Y_j) = np_jp_m[1+(n-1)E(1/\Sigma X)], \qquad j = 1, \ldots, k \tag{73}$$

$$Cov(Y_j,Y_m) = np_jpm[1+(n-1)E(1/\Sigma X)], \; j \neq m . \tag{74}$$

and the correlation of fixed-n counts is as in (34). For independent free counts the conditional argument does not work as before, since probabilities in the conditional multinomial, given Σy (and $\underset{\sim}{x}$), are themselves functions of $\underset{\sim}{x}$ as is also $E(1/\Sigma Y)$.

5. THE LOGNORMAL MODEL

Null Models A, B, and C give variation in population X's and sample counts from "essentially random" situations. Of more interest might be a model capable of describing a wide class of biological possibilities. If, in addition, the model lends itself to standard--that is, normal, statististical theory, the practical treatment of counts could be greatly improved. The lognormal model offers such a possibility, but it remains to be proven applicable in practice.

Suppose the pollen concentrations $X_1, \ldots, X_k$ are such that the random variables $Z_1, \ldots, Z_k$ where $Z_j = \log X_j$, $j = 1, \ldots, k$ are multivariate normal with mean vector $\underset{\sim}{\mu}$ and covariance matrix $\underset{\sim}{\Sigma}$. Then the X's have a multivariate lognormal distribution with parameters $\underset{\sim}{\mu}$ and $\underset{\sim}{\Sigma}$. After the fashion of Aitchison and Brown [1] or Moran [9], moments of the X's are readily obtained:

$$E(X) = e^{\mu_j + (1/2)\sigma_j^2}, \qquad j = 1, \ldots, k, \tag{75}$$

$$\mathrm{Var}(X_j) = E^2(X_j)(e^{\sigma_j^2} - 1), \qquad j = 1, \ldots, k, \tag{76}$$

$$\mathrm{Cov}(X_j, X_m) = E(X_j)E(X_m)\,(e^{\sigma_{jm}}-1), \quad j \neq m. \tag{77}$$

Thus the correlation coefficients are

$$\mathrm{Cor}(X_j, X_m) = (e^{\sigma_{jm}}-1)/[(e^{\sigma_j^2}-1)(e^{\sigma_m^2}-1)]^{1/2}. \tag{78}$$

Of course the corresponding correlation coefficients of the Z's are

$$\mathrm{Cor}(Z_j, Z_m) = \sigma_{jm}/(\sigma_j\,\sigma_m). \tag{79}$$

Corresponding correlations of the Z's and of the X's have the same algebraic sign.

The Z's are multivariate normal; hence

$$E(Z_j|Z_m) = (\sigma_{jm}/\sigma_m^2)\, z_m + [\mu_j-(\sigma_{jm}/\sigma_m^2)\,\mu_m], \tag{80}$$

$$= \beta\, z_m + \tau$$

and the corresponding regression of X_j on X_m can be shown to be

$$E(X_j|X_m) = \delta\, x_m^{\beta}, \tag{81}$$

where $\delta = e^{\tau} + \frac{\sigma_{j\cdot m}^2}{2}$ and $\sigma_{j\cdot m}^2$ is the variance of Z_j given Z_m.

The nonlinear regressions among the X's reduce the utility of their correlation coefficients for expressing relationships.

For example, with $k = 3$, $\sigma_1^2 = \sigma_2^2 = \sigma_3^2 = 9$ and a correlation matrix of

$$\begin{bmatrix} 1 & .8 & .8 \\ & 1 & .8 \\ & & 1 \end{bmatrix}$$

for Z_1, Z_2, Z_3, the correlation matrix of X_1, X_2, X_3 is:

$$\begin{bmatrix} 1 & .17 & .17 \\ & 1 & .17 \\ & & 1 \end{bmatrix} .$$

<u>Distribution of the Pollen Ratios, U's.</u>

If $X_1, \ldots, X_k$ is multivariate lognormal with parameters $\underset{\sim}{\mu}, \underset{\sim}{\Sigma}$ then $\underset{\sim}{U} = \frac{X_1}{X_k}, \ldots, \frac{X_{k-1}}{X_k}$ is multivariate lognormal since $\log U_j = Z_j - Z_k$, $j = 1, \ldots, k-1$. If the pollen concentrations are lognormal, so also are the pollen ratios. Letting $R_j = Z_j - Z_k$, $j = 1, \ldots, k-1$ the moments of the R's are

$$E(R_j) = \mu_j - \mu_k, \qquad j = 1, \ldots, k-1 \tag{82}$$

$$\mathrm{Var}(R_j) = \sigma_j^2 + \sigma_k^2 - 2\sigma_{jk}, \qquad j = 1, \ldots, k-1 \tag{83}$$

$$\mathrm{Cov}(R_j, R_m) = \sigma_{jm} - \sigma_{jk} - \sigma_{mk} + \sigma_k^2, \qquad j \neq m . \tag{84}$$

The R's are the log U's. Hence from (75), (76), and (77) the moments of the U's are

$$E(U_j) = e^{\mu_j - \mu_j + (1/2)(\sigma_j^2 + \sigma_k^2 - 2\sigma_{jk})}, \qquad j = 1, \ldots, k-1 \tag{85}$$

$$\mathrm{Var}(U_j) = E^2(U_j)[e^{\sigma_j^2 + \sigma_k^2 - 2\sigma_{jk}} - 1], \qquad j = 1, \ldots, k-1 \tag{86}$$

$$\mathrm{Cov}(U_j, U_m) = E(U_j)E(U_m)[e^{\sigma_{jm} - \sigma_{jk} - \sigma_{mk} + \sigma_k^2} - 1], \; j \neq 1 . \tag{87}$$

These results could be used immediately in conjunction with (16), (17), and (18) to obtain covariances of the lognormal-compound negative multinomial distribution; however, as just indicated, covariances of log counts would be of more interest under this model.

Given the results in (82) through (87), it is possible to see under various biological situations precisely how information about X's is confounded by (and perhaps in specific cases may be elucidated by) the behavior on the U's.

It is also possible to construct another null model using the following theorem proved elsewhere (Mosimann, [13]).

Let $X_1, \ldots, X_k$ be mutually independent, positive, nondegenerate random variables and suppose that the ratio vector $(U_1, \ldots, U_{k-1})$ is independent of $\left(\prod_{j=1}^{k} X_j\right)^{1/k}$. Then each X_j has a lognormal distribution with the same parameter σ.

Under this null situation the correlation of any two ratios would be

$$Cor(U_j, U_m) = + \frac{e^{\sigma^2} - 1}{e^{2\sigma^2} - 1} \tag{88}$$

Approximate results for proportions have also been found. However unless some intuitive justification can be found for the use of the size variable $\left(\prod_{j=1}^{k} X_j\right)^{1/k}$ to measure total abundance of pollen, then the application of this null situation will be doubtful. An advantage of the lognormal model is that it encompasses a wide range of deviations from this null situation.

REFERENCES

[1] Aitchison, J., and Brown, J. A. C. 1957. The Lognormal Distribution. Cambridge University Press, Cambridge.

[2] Davis, M. B. 1965. A method for determination of absolute pollen frequency. In Handbook of Paleontological Techniques, B. Kummel and D. Raup, eds. W. H. Freeman and Company, San Francisco, 674-86.

[3] ———. 1967. Pollen accumulation rates at Rogers Lake, Connecticut, during late and postglacial time. Rev. Palaeobotan. 2:219-30.

[4] ———. 1969. Climatic changes in southern Connecticut recorded by pollen deposition at Rogers Lake. Ecology. 50:409-22.

[5] Joshi, S. W., and Patil, G. P. 1968. Certain structural properties of the sum-symmetric power series distributions. Technical Reports and Preprints Number 5, The Pennsylvania State University, Department of Statistics.

[6] Livingstone, D. A. 1968. Some interstadial and postglacial pollen diagrams from eastern Canada. Ecological Monographs 38:87-125.

[7] Martin, P. S., and Drew, C. M. 1969. Scanning electron photomicrographs of southwestern pollen grains. Journal of the Arizona Academy of Science 5, No. 3, 147-76.

[8] Martin, P. S., and Mosimann, J. E. 1965. Geochronology of pluvial Lake Cochise Southern Arizona. III Pollen statistics and Pleistocene metastability. American Journal of Science 263:313-58.

[9] Moran, P. A. P. 1968. Introduction to Probability Theory. Clarendon Press, Oxford.

[10] Mosimann, J. E. 1962. On the compound multinomial distribution, the multivariate β distribution and correlations among proportions. Biometrika 49:65-85.

[11] ————. 1963. On the compound negative multinomial distribution and correlations among inversely sampled pollen counts. Biometrika 50:47-54.

[12] ————. 1965. Statistical methods for the pollen analyst: multinomial and negative multinomial techniques. In Handbook of Paleontological Techniques. B. Kummel and D. Raup, eds. W. H. Freeman and Company, San Francisco, 636-73.

[13] Mosimann, J. E. Size allometry: size and shape variables with characterizations of the lognormal and generalized gamma distributions. In press. Jour. of the Amer. Stat. Assn.

[14] Patil, G. P. 1963. On multivariate generalized power series distribution and its application to the multinomial and negative multinomial. In Classical and Contagious Discrete Distributions, G. P. Patil, ed. Pergamon Press, New York.

[15] ————. 1968. On sampling with replacement from populations with multiple characters. Sankhyā. Ser. B 30:355-66.

[16] Pettigrew, H. M., and Mohler, W. C. 1967. A rapid test for the Poisson distribution using the range. Biometrics 23:685-92.

[17] Tauber, H. 1965. Differential pollen dispersion and the interpretation of pollen diagrams. Geological Survey of Denmark. II. Series No. 89, 1-69.

[18] ————. 1967a. Differential pollen dispersion and filtration. In Quaternary Paleoecology E. J. Cushing and H. E. Wright, Jr., eds. Yale University Press, New Haven, 131-41.

[19] ————. 1967b. Investigations of the mode of pollen transfer in forested areas. Rev. Palaeobotan. Palynol. 3, 277-86.

[20] Tiao, G. C., and Cuttman, I. 1965. The inverted Dirichlet distribution with applications. Journal of the American Statistical Association 60:793-805.

[21] Whitehead, D. R. and Tan, K. W. 1969. Modern vegetation and pollen rain in Bladen County, North Carolina. Ecology 50:235-48.

[22] Wright, H. E., Jr. 1968. The roles of pine and spruce in the forest history of Minnesota and adjacent areas. Ecology 49:937-55.

Table 1: Correlation Coefficients associated with various null models. X's preserved pollen concentrations; P's, preserved pollen proportions; U's preserved pollen ratios. For brevity the following notation is used in the table: $E(X_j) = x_j$, $E(\Sigma X_j) = \Sigma x$, $E(P_j) = p_j$, $E(U_j) = u_j$; $j = 1, \ldots, k$ (α is the scale parameter of the gamma distribution in equation (35) of the text).

Null-A	Fixed - n counts (multinomial)	
Null-A	Proportions formed from free counts	
Null-B	Fixed - n counts (Dirichlet (β) compound multinomial)	
Null-B	Proportions formed from free counts	$-\sqrt{\frac{p_j p_m}{(1-p_j)(1-p_m)}} = -\sqrt{\frac{x_j x_m}{[(\Sigma x)-x_j][(\Sigma x)-x_m]}}$
Null-B	Preserved pollen proportions (P's) (Dirichlet)	
Null-C	Fixed - n counts	
Null-C	Preserved pollen proportions (P's)	
Null-A	Inverse counts (negative multinomial)	$+\sqrt{\frac{u_j u_m}{(1+u_j)(1+u_m)}} = +\sqrt{\frac{x_j x_m}{(x_k+x_j)(x_k+x_m)}}$
Null-B	Inverse counts (Dirichlet (β) compound negative multinomial)	$+\sqrt{\frac{u_j u_m}{(1+u_j)(1+u_m)}} = +\sqrt{\frac{x_j x_m}{(x_k+x_j-\frac{1}{\alpha})(x_k+x_m-\frac{1}{\alpha})}}$
Null-B	Preserved pollen ratios (U's) (Inverted Dirichlet)	

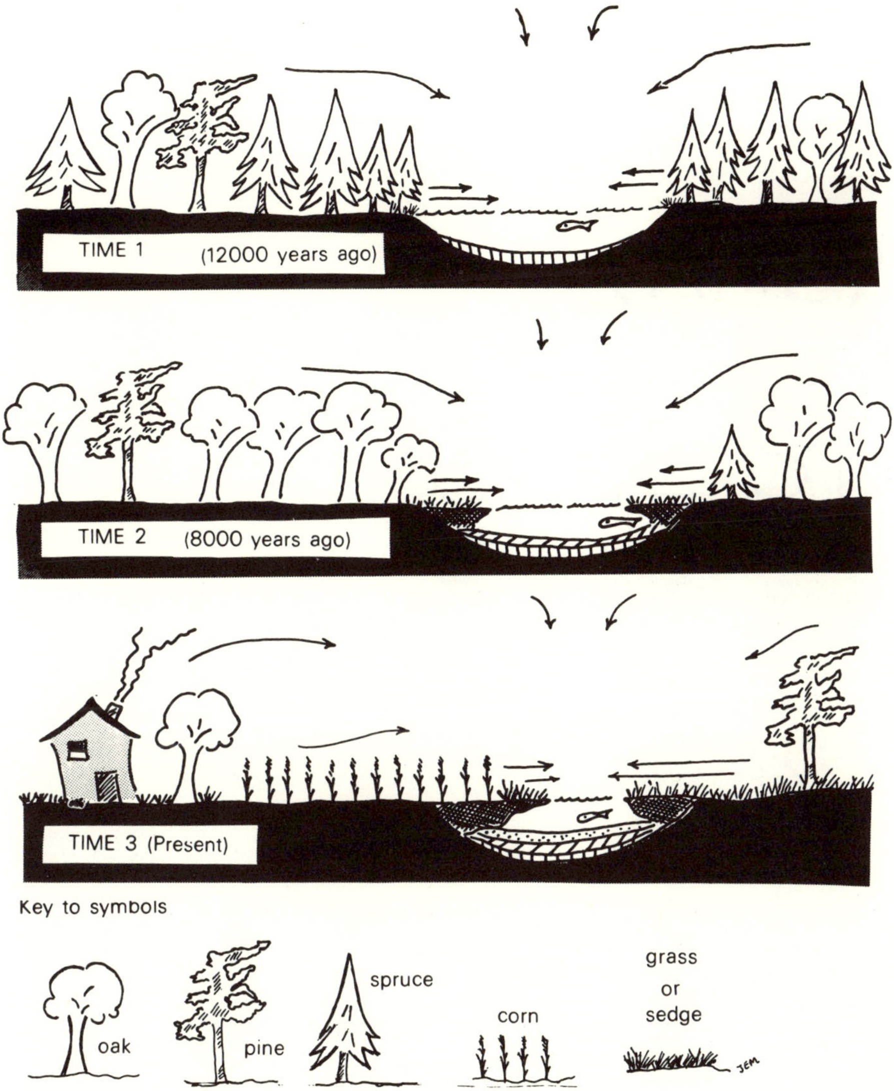

Figure 1. Hypothetical illustration of vegetation change through time. Pollen from the vegetation blows into the bog-lake and is preserved along with accumulating sediment. Changes in kinds of pollen preserved in the sedimentary sequence reflect changes in vegetation (and possibly climate) during the time of the deposition of the sediment.

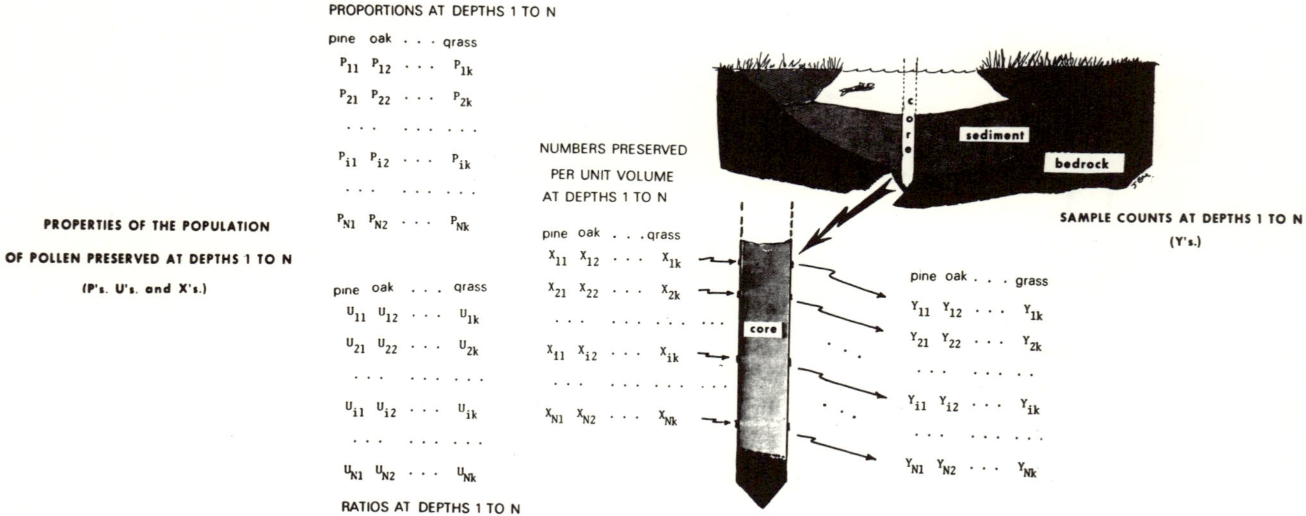

Figure 2. Notation used throughout this paper is illustrated. The concentrations (numbers per unit volume) of preserved pollen at a given depth in the core are denoted by X's; the concentrations at the ith depth being $(X_{i1},\ldots,X_{ik}) = \underset{\sim}{X}$. The proportions of preserved pollen at the ith depth are denoted by $(P_{i1},\ldots,P_{ik}) = \underset{\sim}{P}$ where $P_{ij} = X_{ij} \Big/ \sum_{j=1}^{k} X_{ij}$. Ratios of preserved pollen at the ith depth are denoted by $(U_{i1},\ldots,U_{ik}) = \underset{\sim}{U}$. Where $U_{ij} = X_{ij}/X_{ik}$ with $U_{ik} = 1$. X's, P's and U's are characteristics of the population of preserved pollen at the ith depth and may not be considered values of random variables.

Sample counts at the ith depth are denoted by $Y_{i1},\ldots,Y_{ik}$. The distribution of the sample counts at the ith depth will be determined by the method of counting. X's, P's, or U's will be parameters of the distribution for the counts.

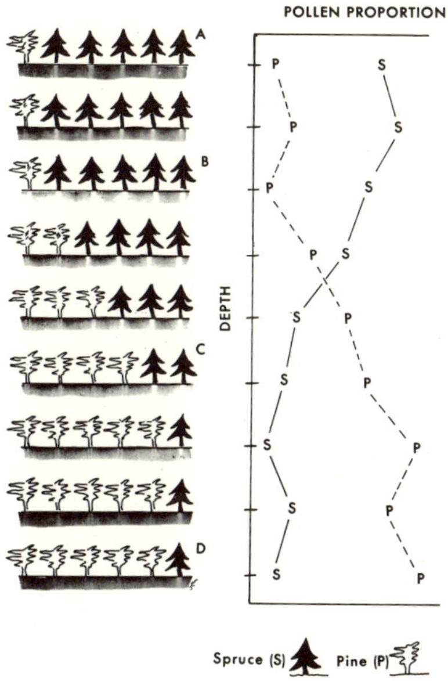

Figure 3. An illustration of various possible models for depth variation described in Section 3 of the text.

An Hypothetical situation is depicted illustrating a gradual transition from pine-dominated vegetation (bottom) to spruce-dominated vegetation (top). The figures at the left illustrate the relative composition of the two types. At the right, proportions formed from sample pollen counts are plotted (the two proportions do not total to one).

A single trend is indicated and clearly over the entire profile the vegetation history is such that the x's would not be independently determined from level to level. Thus Model I, (see Section 3) describing the history in this core, or perhaps Model II(b) regarding vegetation history itself as one sample from a time series would be appropriate.

From depths A to B the proportions of spruce and pine in the vegetation do not change. Within this time pollen production still might vary. The depth to depth variation in pollen might be more or less independent for level to level and one might have Model II(a) applicable. In fact Null Model A or B (Section 4) might be applicable although these both involve adding detail to the situation portrayed.

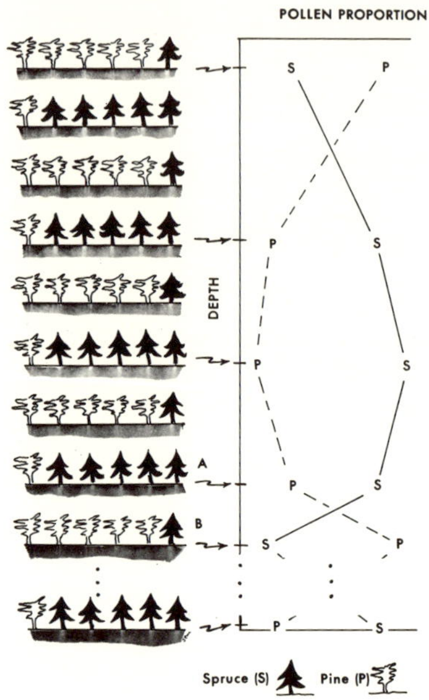

Figure 4. An illustration of various possible models for depth variation described in Section 3 of the text.

Here the profile at left represents a long period of time with a number of pine-dominated to spruce-dominated vegetation transitions (and vice versa). The sampling in terms of depths is incomplete so that entire transitions and vegetational events are missed. Because of the incompleteness of the sampling the variation in X's might be independent from level to level for practical purposes. Model II(a) may be applicable here.

From level B to A there occurs a single transition such as illustrated in detail in Figure 3.

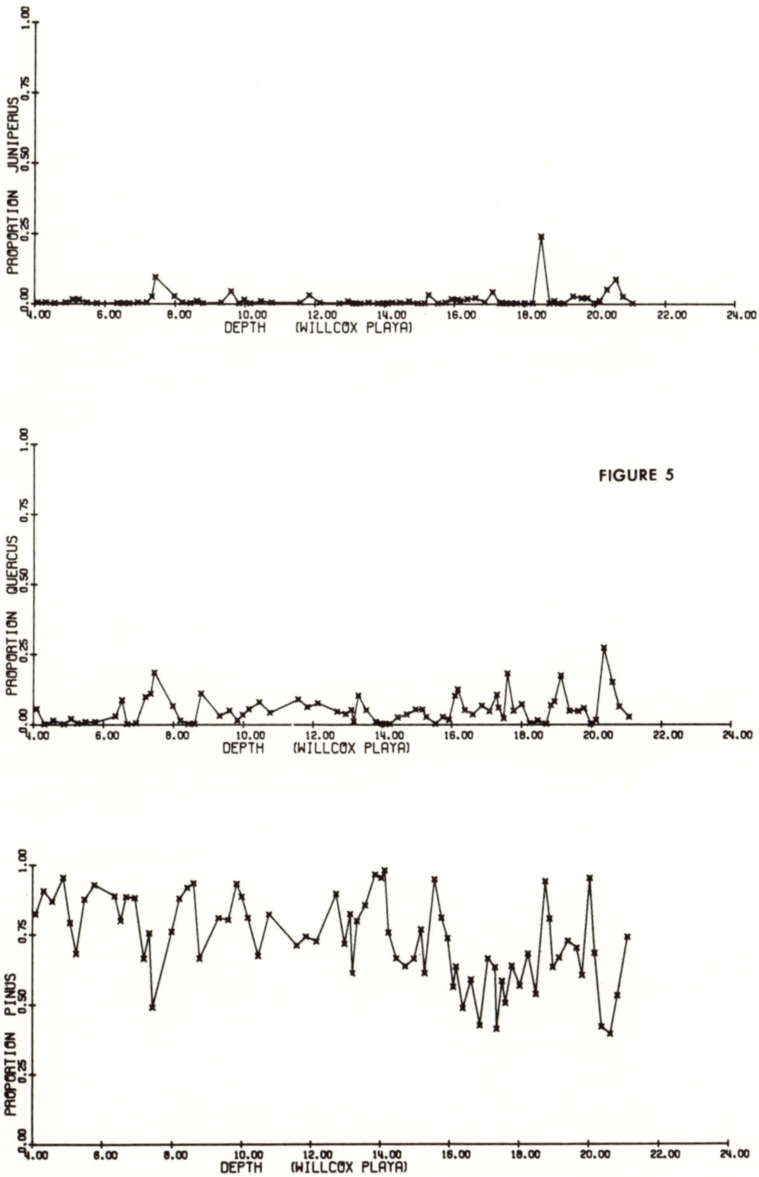

Figure 5. Illustration of depth-variation models.

Depicted are pollen proportions from a portion (Divisions III and IV of Martin and Mosimann [8] of a long core taken from the Willcox Playa in southern Arizona. Proportions are formed from fixed - n 200 grain counts. *Pinus*, *Quercus* and *Juniperus* denote pine, oak, and Juniper-type pollen respectively. Depth is in meters. The spacing of samples is too wide to give smooth transitions such as illustrated in Figure 3 and the situation is presumably more similar to Figure 4. The core is of considerable age and depths from 1.9 to 3.4 meters (not shown) are correlated by pollen and radiocarbon dating with the Wisconsin glacial maximum (Ca 24,000 years before present). All the counts shown here are prior to this in time.

Although there is some time trend, the independent levels, Model II(a), can be used for practical purposes, particularly if the graphs shown are divided into Division III (depths 4 to 14.1) and Division IV (depths greater than 14.1). Correlograms of pine are given in the work cited. For an analysis of the single pine peak indicated at 19 meters here (see Figure 6).

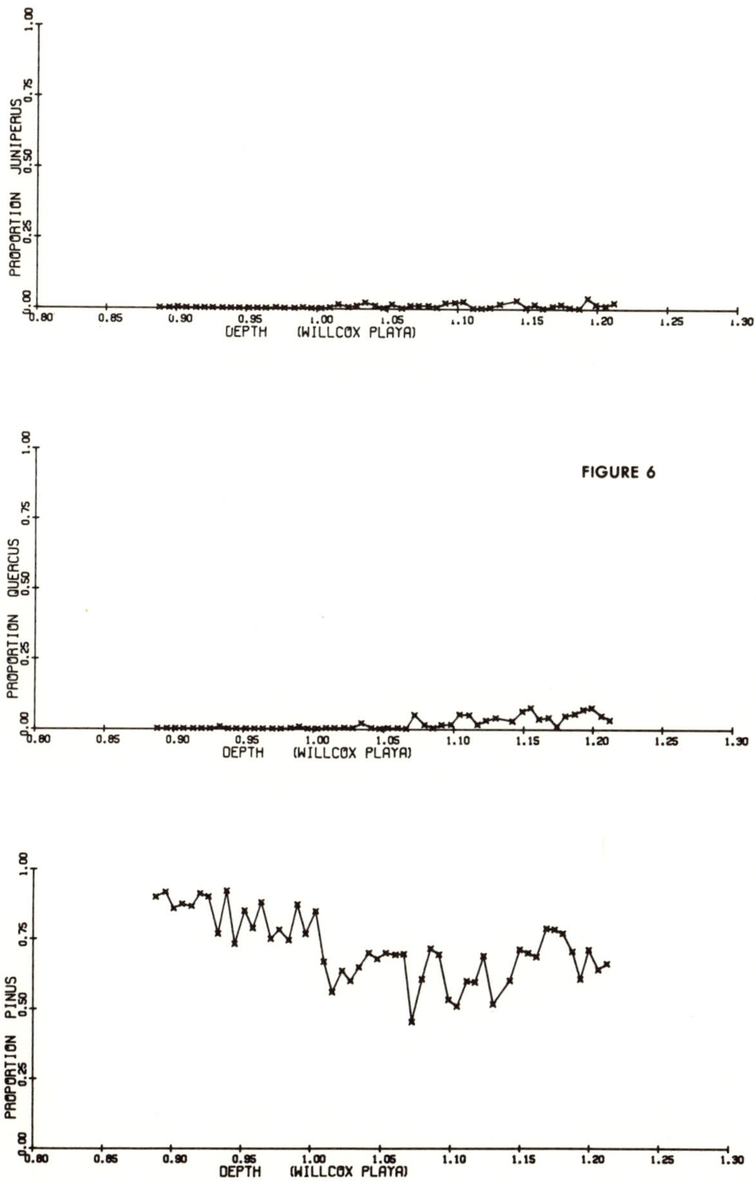

Figure 6. Illustration of depth variation models.

Close interval sampling of the Willcox core from 19.9 to 20.2 meters depth: See also Figure 5. Depths here are measured from 19 meters. The counts represent those from core slug 96 of Martin and Mosimann [8].

There is a gradual decrease of oak (=Quercus) and Juniper-type pollen from 20.2 to 20 meters with an accompanying increase in pine pollen at 20 meters continuing to 19.9 meters. The situation is thought to represent a shift in vegetation from oak-juniper woodland growing near lake-level with ponderosa pine parkland growing at just higher altitudes (20.2 to 20 meters) to a cooler-moister time in which pine parkland bordered the lake itself (20 to 19.9 meters).

The situation is analogous to the model of a single transition portrayed in Figure 3. Model I might be most usefully applied to the whole of Figure 6 because of the time trend in the p's (and presumably x's) but model II(a) with its independent determinations of p's and x's might be applied within the zone representing the time when pine parkland bordered the lake. For an application of model II(a) and also correlograms for pine see the work cited above. Thus during the time when the vegetation was not changing appreciably, pollen production and delivery to the site of deposition might vary independently from depth to depth due to various causes; variation in humidity, wind-direction, flowering success, etc.

STRUCTURE BY SAMPLING IN THE GEOSCIENCES

JOHN C. GRIFFITHS*
CHARLES W. ONDRICK**
University of Kansas
State Geological Survey

SUMMARY

An algorithm for sampling geological populations with a view to estimating proportions of elements of different kinds (i.e., determining mineral composition) is proposed, and examples of applying this algorithm to counts of individual elements and aggregates under the microscope are discussed. When the procedure is applied to the distribution of non-renewable natural resources in a region, the analysis is similar to the study of aggregates under the microscope except for a scale factor.

Simulations of situations resembling real-world exploration for natural resources suggest that the estimates of occurrence of resources, based on conventional practice, may be biased; areas potentially rich in resources may be rejected if the bias revealed by the simulations actually exists.

*Visiting Research Associate, Kansas Geological Survey, 1968-69

**Pan American Postdoctoral Fellow, Kansas Geological Survey, 1969-70

1. INTRODUCTION

The structure of a population may be defined as the pattern of variation of an element (or variate) in the population; therefore, a population, an element, and the variate of interest are given (see Figure 1). The next step is to formulate a model which will generate a frequency distribution of the variate in the population; this frequency distribution model is usually selected heuristically, and estimates of the parameters of the model then become the objective of the experiment.

It is first necessary to define a sample of n elements where n = 1,2, etc.; samples are selected from the population by using a randomizing device, for example, a random number table (RAND Corporation, "A Million Random Digits" [14]) or a computer-based random number generator. The variate is counted (or measured) on the elements of the sample, and an observed frequency distribution is constructed together with its statistical estimators. These outcomes, the observed frequency distribution, and the statistics, are a function of the element arrangement in the population, the sampling arrangement used to collect the data, and their mutual interaction (see Figure 1).

The statistical estimators are compared with the model parameters by moment analysis, e.g., the Pearsonian b_i estimators, by chi-square and/or analysis of variance, etc. Outcomes of the testing yield decisions on whether or not the estimators are biased; that is, they are either significantly different from the parameter values or there is no evidence to reject the hypothesis.

If the null hypothesis is not rejected, then the arrangement of the variate in the population is homogeneously random; in a geological sense the population, usually a rock-body of some kind, is massive with regard to the structural arrangement of this variate. Essentially, then, these are random samples from some homogeneous population, and the sampling arrangement is independent of (orthogonal to) the pattern of arrangment of the element (variate) in the population. There may be a pattern or structure in the population, but the sampling arrangement does not interact with it, and hence the estimators are stable.

On the other hand, if the hypothesis is rejected on the basis of one or more of the significance tests, then it may be stated that the sampling arrangment interacts with the pattern of

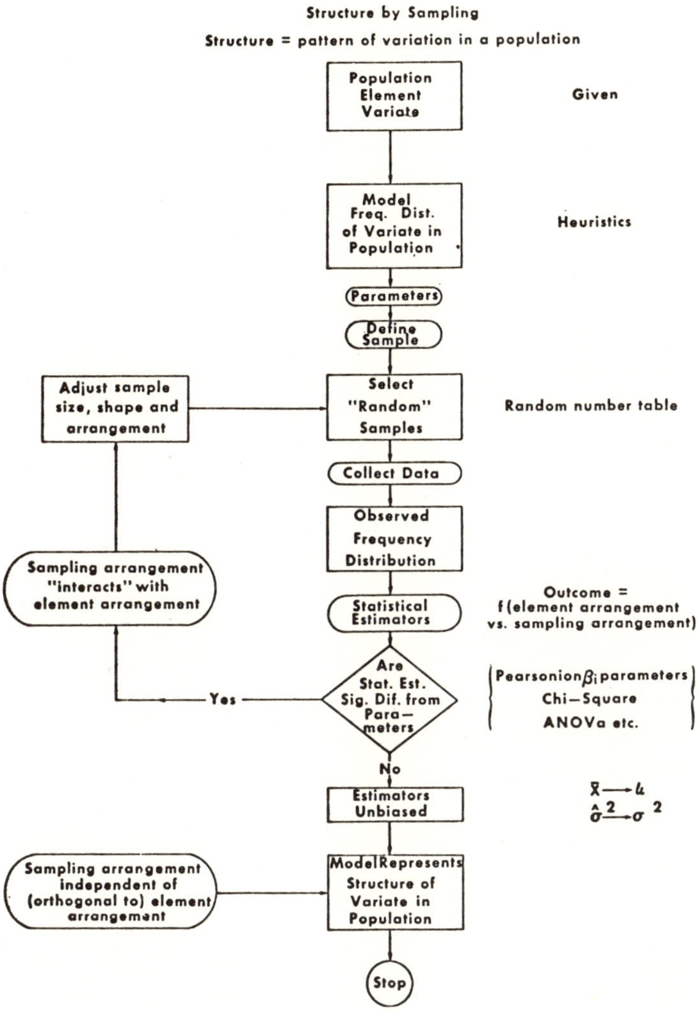

Fig. 1. Structure by sampling.

variation of the variate in the population; the estimators are biased. In general, the sample may then be redefined in terms of its size, shape and arrangement, additional data collected, and the testing repeated as shown by the cycle in Figure 1. Ultimately the significance tests fail, and the alternative conclusions hold.

It seems evident that, by appropriate choice of sample size, shape and arrangement, many different frequency distributions may be produced from a single structural arrangement of a variate in a population; the role of the model is to establish a calibration standard for deciding among the alternatives. The presence of interactions among sampling arrangement and structural arrangement and its removal by suitable adjustments of sampling arrangement serves to define the structure of the variate.

Since a channel sample is a self-weighting random sample across the structure (for example, across a layered structure in a sedimentary rock), it leads to unbiased estimators, and the sampling is orthogonal to the structural arrangement; if this kind of sampling arrangement is adopted, then no structural arrangement of the variate within the population is obtained--the structure is treated as if it is massive. In order to define the structure it is necessary to sample in such a way as to reflect structure, and therefore two-dimensional sampling is necessary for a surface and three-dimensional sampling is required to define a structure in three dimensions. In these cases, if an interaction occurs between the dimensions (directions), the sampling arrangement is oblique to the structure, and the estimators are usually biased. When, upon suitable adjustment of the sampling, the interaction disappears, the sampling arrangement is orthogonal to the structural arrangement, and variation within and between rows and columns leads to a definition of structure of the variate. Examples of the application of this algorithm to geological populations are described in Griffiths [9], Griffiths and Ondrick [11], Bennion and Griffiths [1], and Dahlberg and Griffiths [3].

2. DEFINING STRUCTURE BY SAMPLING GEOLOGICAL POPULATIONS

Two different kinds of procedures will be considered as the basis for illustrative examples. The first applies to studies of different kinds performed under the microscope, and the second is concerned with field studies using a map as the basic frame of reference. Actually, both kinds are equivalent in terms of analysis, except for a scale factor, and the sampling procedures are invariant to scale changes from fractions of a micron (electron microscopy, for example) to many miles (regional geological maps of the earth's surface); that is, the sampling procedures appear to be invariant to change of scale from 1 to 10^{14} or thereabouts.

2.1 Microscopic Techniques Applied to Individual Elements

A number of experimental procedures used in geological problems lead to mounts of individual elements upon which counts are made to determine the proportion of elements of different kinds in a population; for example, accessory heavy mineral analysis, pollen and microforaminiferal counts, etc. In all these cases the structural arrangement of the element in the mount is the outcome of the experimental procedure and so is, theoretically at least, under control of the experimenter. In order to obtain unbiased estimators of the parameters it is necessary to obtain random samples; that is, the sampling arrangement should be independent of the structural arrangment of the element in the mount. The models are binomial if the element forms between 5 and 95 percent of the population and Poisson if less than 5 or more than 95 percent.

An example which exhibits the procedure for accessory mineral analysis is summarized in Figure 2. The left-hand column illustrates the separate steps involved, all of which are, in effect, subsampling procedures and each of which introduces its own contribution to subsampling error (e.g., Krumbein and Rasmussen, [12]). The ultimate step leads to a mount in a glass slide and the estimate of proportion, $\bar{X}$, is determined by counting traverses across the slide. If the elements (in this example, mineral grains) are uniformly irregular across the slide (case 1A, Figure 2), the estimator ($\bar{X}$ = 30.0) is unbiased, and the sampling arrangement does not interact with the arrangement of grains in the slide; on the other hand, if the grains are clustered (case 1B, Figure 2), the sampling arrangement interacts with the element arrangment, and the estimate ($\bar{X}$ = 30.0) is biased.

In the second case it is necessary either to make another mount which will resemble case 1A, or to adjust the sampling, say, "perpendicular" to the clustering, to obtain unbiased estimators. It is generally possible to find a suitable sampling arrangement which will achieve the desirable objective and Fisher's Dispersion Index may be used as a criterion for determining the suitability of the sampling arrangement (Fisher [6], page 58; Griffiths [9], pp. 286, 295).

What is not generally recognized is that when this desirable outcome has been accomplished, the result is that we know the proportion of the element in the mount. The procedure must be repeated at each stage to ensure that the estimator is an unbiased estimator (of suitable precision) to specify the proportion of the element in the rock-body, i.e., the population of geological interest. All the separate steps involved in case 1 (Figure 2) are aimed at increasing the proportion of the element in each succedding residue to facilitate the ultimate count. In the case of an accessory heavy mineral, this means that the element (mineral grain), which was initially a very rare event (i.e., it occurs in very low proportion in the specimen), is now more common in the mount. In other words, the entire procedure in case 1 (Figure 2) is equivalent to transforming the expected model frequency distribution from a Poisson to a binomial; it would appear much more efficient and simpler to use the Poisson model as in case 2 (Figure 2), and the outcome, an unbiased estimate of the proportion of the element in the population, would be very much easier to achieve.[1] Furthermore, in case 2 the subsampling error is essentially removed by commencing with a sample of suitable size so that no subsampling is introduced below the specimen level.

2.2 Microscopic Techniques Applied to Aggregates.

An entirely different set of procedures gives rise to thin sections of rocks, which are also studied under the microscope (Chayes, [2]); however, in this case, the structure of the element in the population is not under the control of the experimenter. The structure of the material is now of considerable interest to the analyst, and it is usually necessary to

---

[1] It is assumed here that the objective is to estimate the proportion of the element in the population; if the objective is to identify some rare constituent, then the problem of estimation does not arise, and the various steps in the technique are quite appropriate.

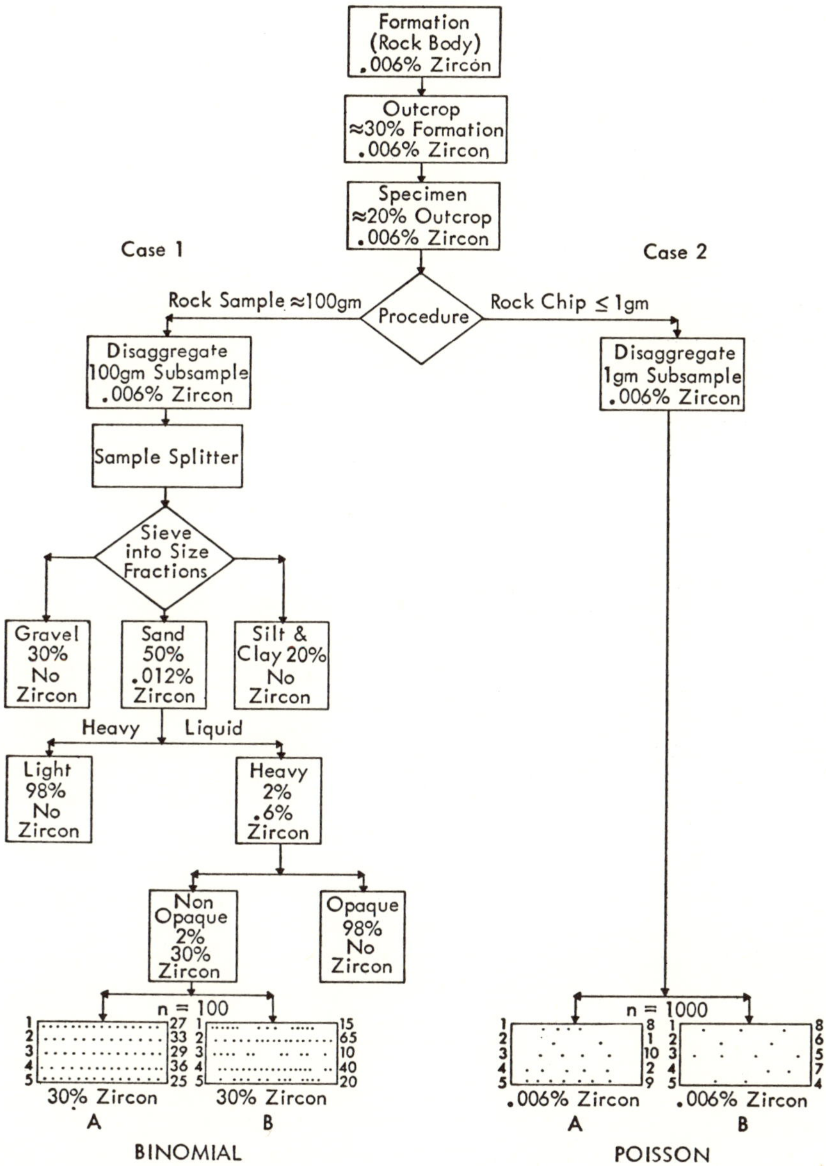

Fig. 2. Experimental procedure determining the proportion of elements having accessory heavy minerals.

define it carefully (Wood and Griffiths [17]). Here the only alternative is to find a sampling pattern which will yield unbiased estimators of population parameters and which in some sense will reflect the pattern of variation of the element in the population.

The patterns which tend to occur most frequently in sedimentary rocks are layered (stratified), patchy, or massive (patternless). If it is known that the pattern is massive, then most sampling arrangments are independent of structural arrangment, and the estimators are unbiased. However, when it is not known that the structure is massive, the establishment of absence of pattern is extremely difficult, and the sampling arrangment is critical.

It is convenient to consider a layered or stratified pattern first, and the sample size and shape should then fulfill the following constraints:

a. The samples must be large enough to include the largest element.

b. The samples must be small enough to fit within the layers. Counts in small sets, of sample size less than 10 and usually in the range 4-6, are appropriate; furthermore, the arrangment of samples is important, and for reconnaissance a square or rectangular grid is most useful. In most geological cases the observer believes he can recognize the layered structure, and then it is advisable to arrange the grid so that one dimension is parallel and the other perpendicular to the layering.

In the most favorable case the samples, labelled for convenience as columns, are arranged perpendicular to the structure; they are channel samples which are random, self-weighting samples yielding unbiased estimators of the population mean and variance. If the samples arranged paralled to the structure (in rows) are then adjusted until they yield estimators not significantly different from the channel (column) samples, they also yield unbiased estimators, and the weighting required to bring them into correspondence with the channel estimators represents the proportions of the different layers in the population (Griffiths and Ondrick [11]).

Here the presence or absence of interaction between rows and columns is an important feature of the design; if significant interaction is present, then the grid arrangment is oblique to the structural arrangement; and, if the grid is square or rectangular, it may be rotated until the interaction becomes nonsignificant. This rotation yields the direction of layering, and a minimax criterion[2] may be used to decide when the layered structure is adequately defined. This analysis may be extended to three (or more) dimensions with similar outcomes ([1] and [3]).

2.3 Techniques Applied to Natural Resources; Use of Maps

When the population is defined as a region and the elements are nonrenewable natural resources, such as oil and gas fields, ore deposits, or sands and gravels, the procedure for analysis is similar to analyzing aggregates under the microscope except for a scale factor. The basic frame of reference for the population is a map of the region, upon which is imposed the apparent shape of the elements, that is, the shape of the resource bodies as projected on the surface of the earth represented by the map. By superimposing an appropriate grid on the map the probability of a hit on a resource may be estimated (Griffiths [7], [8], and [10]; Savinski [15]; and Drew, [4]). The expected frequency distributions are Poisson if the resources are relatively rare ($P < 0.05$) and binomial if the resources are "common" ($0.05 < P < 0.95$); in most cases the frequency distributions observed are more closely approximated by a negative binomial,[3] i.e., the resources tend to be clustered ([7], [8], [10]). This outcome may be interpreted to imply that natural resources tend to be clustered for genetic reasons, or it may be a reflection of the prevailing philosophy of exploration, namely, that when a resource is found, more may be found nearby. This ambiguity may be illustrated by the following tests, which approximate some of the general practices used in exploration for natural resources.

Suppose the population, i.e., region of interest, is 300 miles square and that the desirable resource is distributed at random according to a Poisson probability with $p = 0.005$. The element of area is a square mile and each square mile either contains a resource or does not; the sample size is $n = 100$ square miles

---

[2] Maximum variation between row means and minimum variation within rows.

[3] Computer programs for fitting observed data to these three distributions and for testing the fit by chi-square are described by Ondrick and Griffiths [13].

and there are N = 900 of these in the region. These conditions may be simulated by generating a distribution of resources by means of a random number generator on a computer and plotting the response points (i.e., squares with a resource) on a map, as in Figure 3. For geological or economic reasons the subregions marked "covered" are inaccessible, and therefore exploration is confined to the "uncovered" areas. A likely example would be a search for gold by examining surface outcrops, and, in the region of interest, the areas covered underly alluvium and gravel terraces and so are inaccessible to surface examination. Alternately, company leasing is confined, by lack of availablility of leased area, to the uncovered parts of the region. The question arises: what is now the frequency distribution of resource in the region?

The frequency distributions, observed, expected Poisson and negative binomial, of the natural resource in the entire region before covering is listed in columns 2 to 4 of Table 1a. The results indicate that the fit to a Poisson is excellent, as would be expected. When three continuous areas are covered (21 percent covered area), with the coverings resembling river valley alluvium as in Figure 3, the Poisson model is rejected and the negative binomial accepted. Results of this simulation are recorded in columns 5 to 7, Table 1a.

In the second simulation fifty percent of the area is covered (Figure 4), and the resulting frequency distributions for the whole area (covered and uncovered) are recorded in columns 8 to 10, Table 1a. Again the Poisson is rejected and the negative binomial accepted. On the other hand, if the frequency distributions of the resources in the covered and uncovered areas are separately considered, they remain Poisson even when the sampling is as nonrandom as those represented by the covered and uncovered areas (see Table 1b).

In another series of simulations a 10,000 square mile area (N = 100; n = 100) is investigated where the probability, still Poisson, for encountering a resource target, equals 0.01 (see Figure 5). Two simulated coverings are placed on the original map; simulation 1 (Figure 7) is "kidney-shaped" and simulation 2 (Figure 5) is more regular in areal extent. Frequency histograms for the observed, expected Poisson and negative binomials are illustrated in Figure 6. The corresponding frequencies are listed

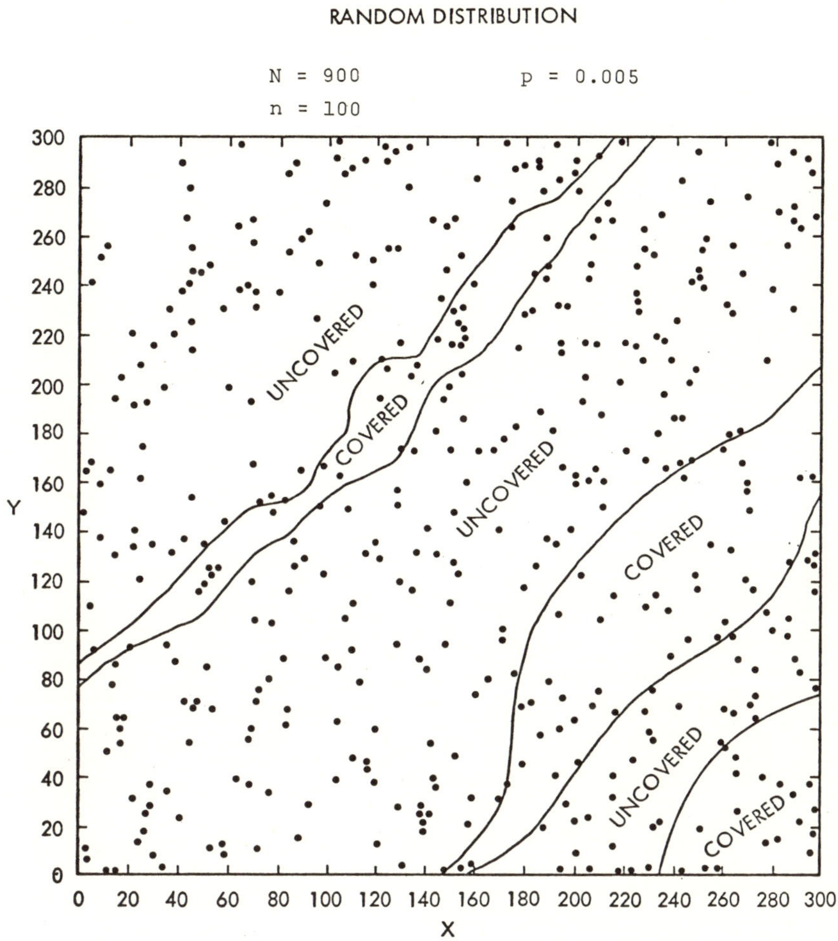

Fig. 3. Computer simulation of a (300m. X 300m.) region with 21% area covered. Points showing resources.

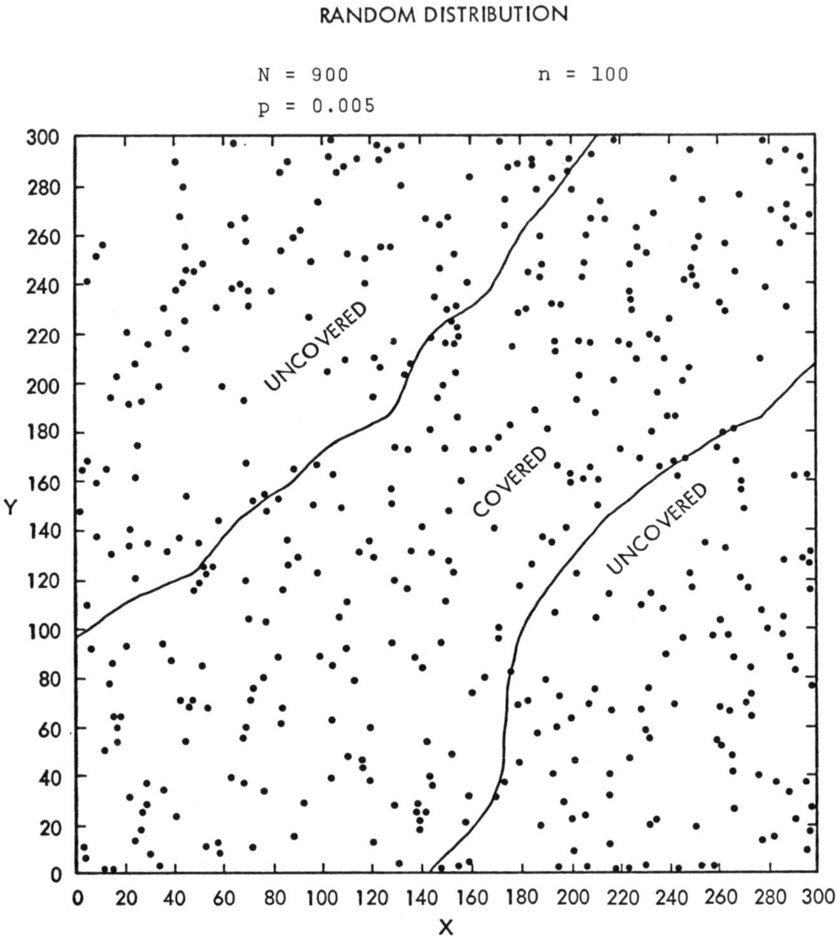

Fig. 4. Computer simulation of a (300m. X 300m.) region with 50% area covered. Points showing resources.

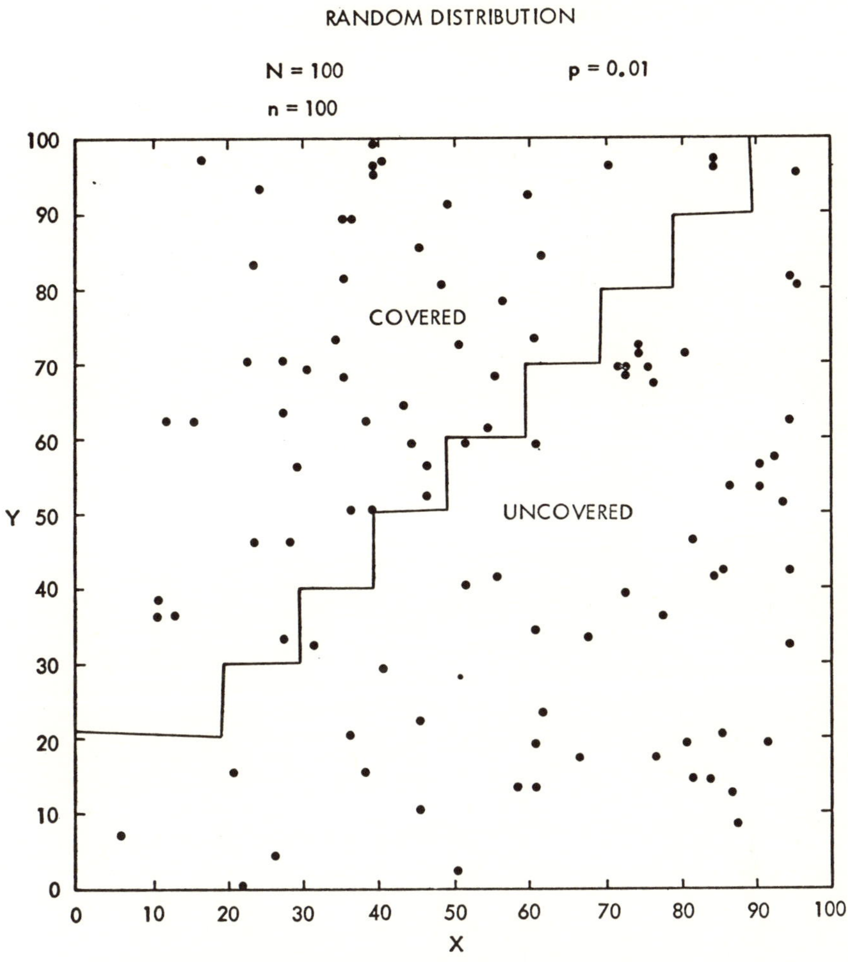

Fig. 5. Computer simulation (2) of a (100m. X 100m.) region with 54% area (regular shape) covered. Points showing the resources.

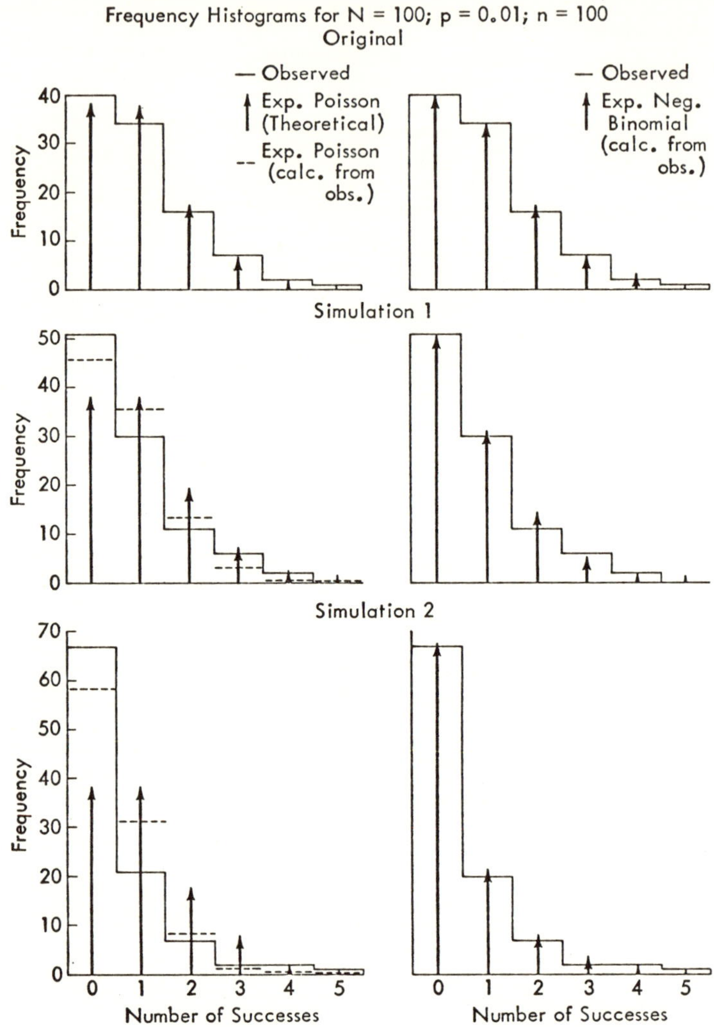

Fig. 6. Frequency histograms for N = 100; P = 0.01; n = 100.

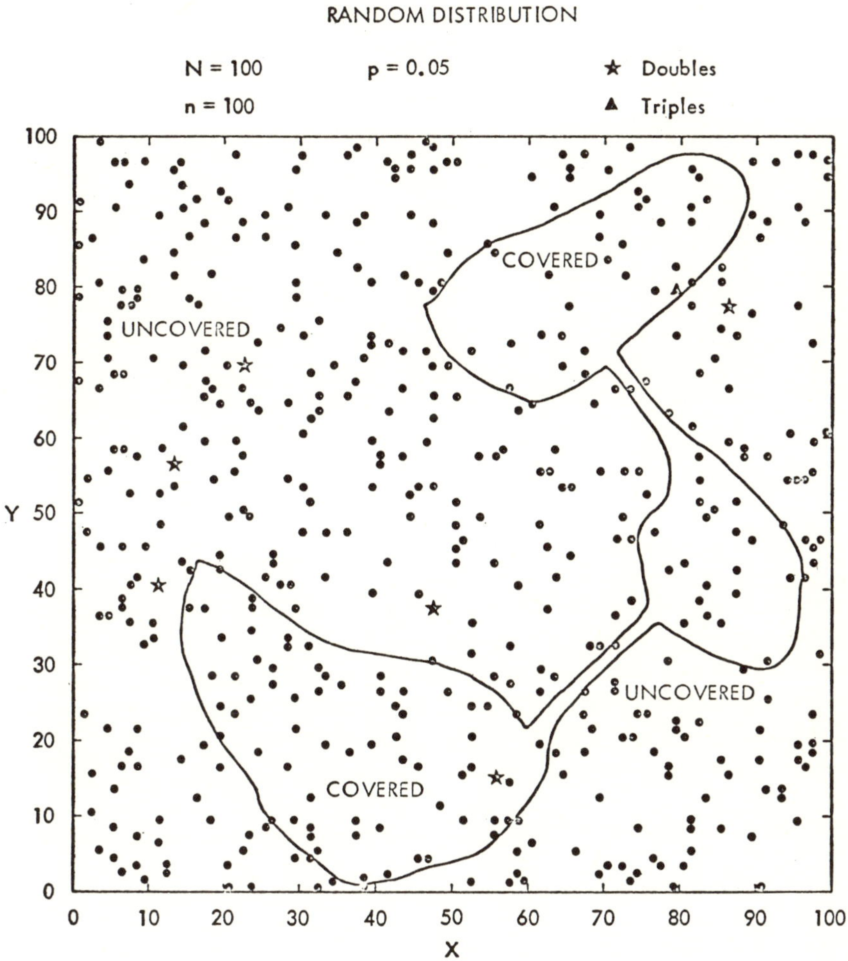

Fig. 7. Computer simulation (1) of a (100m. X 100m.) region with 18% area (kidney-shaped) covered. Points showing resources.

in Table 2a, columns 2 through 4 for the entire area, when uncovered, and in columns 5 through 7 and 8 through 10 for simulations 1 and 2, respectively. It is to be pointed out that, here again, as in the case of the N = 300 area above, there is a tendency to develop clustering when areas are inaccessible, i.e., covered. In this case, when the covered and uncovered areas are treated separately, there is a tendency in the first simulation to reduce the fit of the Poisson; i.e., there is a tendency towards clustering.

Using the same areal dimension (N = 100; n = 100) with p = 0.05 and the same two types of covering, kidney-shaped and regular, the frequency distributions for the entire region are illustrated by the histograms of Figure 8. The corresponding frequencies are listed in Table 3a. The effect of the coverings is to reject both Poisson and negative binomial distributions. When the covered and uncovered areas are individually examined, the favorable fit to the Poisson is maintained, although it would appear that the negative binomial yields a lower chi-square value, irrespective of degrees of freedom available.

It seems evident from Figure 8 and Table 3a that the observed distributions with coverings yield more successes more frequently in the 6, 7, and 8 classes than would be expected from the Poisson and negative binomial distributions with corresponding mean values.

A number of other simulations in which the size of region and the probabilities of success were varied yielded similar results; in general it seems that any attempt to constrain the area has a similar tendency to induce clustering in the distributions. We conclude that if natural resources are, in fact, distributed according to a constant probability model, such as Poisson, then any rejection of large and continuous parts of the population will usually tend to result in a frequency distribution which shows some degree of clustering. In the case of the whole region, when areas are rejected, they are subsequently considered as areas of zero success and so they inflate the zero class; this in itself tends to generate a negative binomial from a Poisson distribution, essentially by generating a mixed distribution from a homogeneous one (Student [16]).

These simulation experiments do not suffice to "prove" that the clustering of natural resources, which is an observed outcome from conventional exploration programs, is not a true reflection

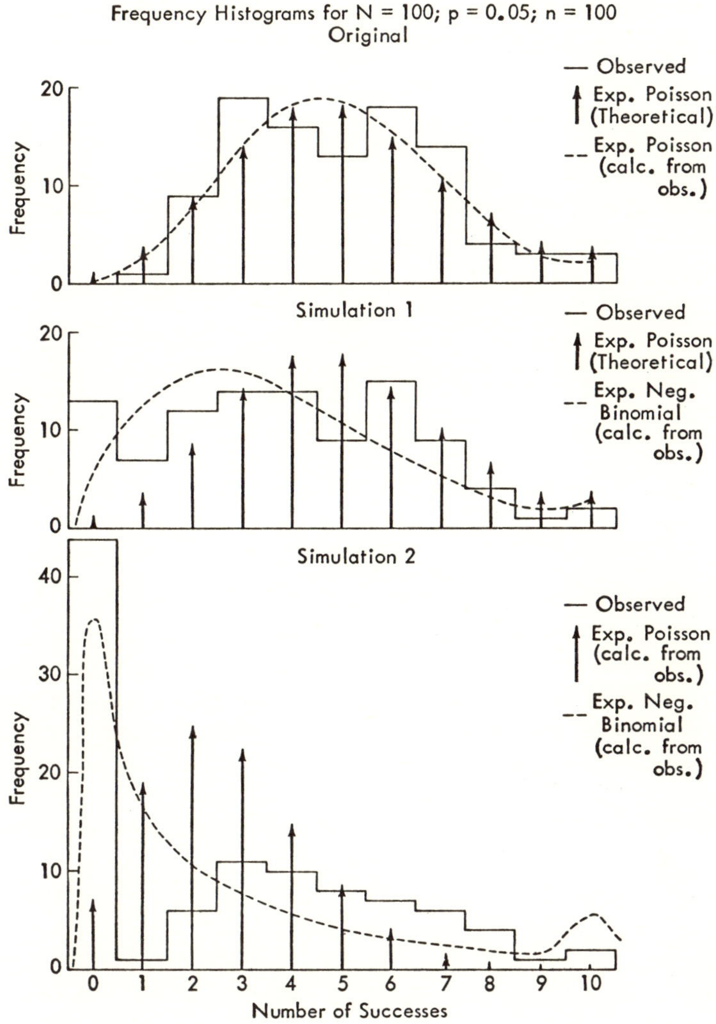

Fig. 8. Frequency histograms for N = 100; p = 0.05; n = 100.

of the distribution of natural resources; but they show that such an observed result may indeed be misleading and may arise from interaction between (nonrandom) sampling and resource arrangements. If natural resources do approximate a constant probability model, such as the Poisson distribution, then present exploration strategy appears to be rejecting areas which may be as remunerative as those they are accepting; the only way to resolve the dilemma is by extended sampling of the entire region.

3. ACKNOWLEDGMENTS

We would like to thank the State Geological Survey, University of Kansas, for providing the support for much of this research, and the Computation Center, University of Kansas, for advice in generating the random numbers for the coordinates in Figures 3, 4, 5, and 7, and also for the service facilities for the computation.

REFERENCES

[1] Bennion, D. W. and Griffiths, J. C. 1966. A stochastic model for predicting variations in reservoir rock properties. Jour. Soc. Petrol. Engr.: v. 6, pp. 9-16.

[2] Chayes, F. 1956. Petrographic Modal Analysis. John Wiley & Sons, Inc., New York, 113 p.

[3] Dahlberg, E. C., and Griffiths, J. C. 1967. Defining gradients in a sample of sedimentary rock. Sedimentology, v. 8, no. 4, pp. 291-309.

[4] Drew, L. J. 1967. Grid-Drilling Exploration and Its Application to the Search for Petroleum, Econ. Geol. 62:698-710.

[5] Fisher, R. A. 1948. Statistical Methods for Research Workers. Oliver and Boyd, Edinburgh (10th ed.), 354 p.

[6] Fisher, R. A., and Yates, F., 1948. Statistical Tables Biological, Agricultural and Medical Research, 3rd ed., Oliver and Boyd Ltd., Edinburgh, 112 p.

[7] Griffiths, J. C. 1962. Frequency Distribution of Some Natural Resource Materials. The Pennsylvania State Univ. Min. Ind. Expt. Sta. Circ. No. 63, pp. 174-98.

[8] ————. 1966. Exploration for Natural Resources. Jour. Op. Res. Soc. Amer. 14:189-209.

[9] ——————. 1967a. Scientific Method in the Analysis of Sediments. McGraw-Hill Book Co., Inc., N.Y., 508 p.

[10] ——————. 1967b. Mathematical Exploration Strategy and Decision Making, Proc. 7th World Petrol. Congr. 11:87-95.

[11] Griffiths, J. C., and Ondrick, C. W. 1968. Sampling a Geological Population (Computer Contribution 30) State Geological Survey, Univ. of Kansas, Lawrence, Kansas, 53 p.

[12] Krumbein, W. C., and Rasmussen, W. C. 1941. The Probable Error of Sampling Beach Sand for Heavy Mineral Analysis. Jour. Sed. Pet. 11:10-20.

[13] Ondrick, C. W., and Griffiths, J. C. 1969. Fortran IV Computer Program for Fitting Observed Count Data to Discrete Distribution Models of Binomial, Poisson and Negative Binomial (Computer Contribution 35) State Geological Survey, Univ. of Kansas, Lawrence, Kansas, 20 p.

[14] RAND Corporation, 1955. A Million Random Digits. The Free Press of Glencoe, New York, 200 p.

[15] Savinski, I. D. 1965. Probability Tables for Locating Elliptical Underground Masses with a Rectangular Grid, Consultants Bureau, N. Y., 110 p.

[16] Student, 1919. An Explanation of Deviations from Poisson's Law in Practice, Biometrika 12:211.

[17] Wood, G. V., and Griffiths, J. C. 1963. Modal Analysis of Three Quartzites and Their Textural Implications, J. Geol. 71:405-21.

Table 1. Simulated frequency distributions of natural resource targets for a population of N = 900; p = 0.005; n = 100.

a. Observed, expected Poisson and negative binomial frequency distributions for the original uncovered area and simulated coverings (covered area counted as no-target responses).

Class	Original Obs.	Original Expected Poisson	Original Expected Neg. Bi.	Simulation 1 Obs.	Simulation 1 Expected Poisson	Simulation 1 Expected Neg. Bi.	Simulation 2 Obs.	Simulation 2 Expected Poisson	Simulation 2 Expected Neg. Bi.
0	547	545.88	544.49	608	896.407	604.692	720	897.304	718.43
1	270	272.94	275.65	220	3.586	229.675	133	2.692	140.13
2	70	68.23	67.74	61	0.007	53.773	38	0.004	31.61
3	13	11.37	10.76	11	0	9.978	9	0	7.45
4	0	1.42	1.24	0	0	1.609	0	0	1.796
5	0	0.15	0.11	0	0	0.273	0	0	0.579
Total	900	899.99	899.99	900	900.000	900.000	900	900.000	899.995
Chi-Square	---	0.0855	2.266	---	---	1.459	---	---	1.726
D.F.	---	2	1	---	0	2	---	0	2
Prob.	---	98-95	70-50	---	Sig.	30-20	---	Sig.	20-10

Table 1. Simulated frequency distributions of natural resource targets for a population of N = 100; p = 0.005; n = 100.

b. Observed, expected Poisson and negative binomial frequency distributions for both covered and uncovered areas in two simulated coverings of the original area.

Simulation 1 21% Area Covered

	Uncovered			Covered		
Class	Observed	Poisson	Neg. Bi.	Observed	Poisson	Neg. Bi.
0	417	417.797	414.372	130	129.655	130.102
1	220	220.979	226.014	50	50.233	49.506
2	61	58.439	58.108	9	9.731	11.378
3	11	11.769	10.499	2	1.378	
Total	709	708.984	708.993	191	190.997	190.986
Chi-Square		0.168	0.344		0.337	0.017
D.F.		2	2		2	1
Prob.		95-90	90-80		90-80	90-80

Simulation 2 50% Area Covered

	Uncovered			Covered		
Class	Observed	Poisson	Neg. Bi.	Observed	Poisson	Neg. Bi.
0	272	266.955	270.547	275	279.707	273.341
1	131	139.410	134.147	139	133.016	142.958
2	37	36.401	36.647	33	31.628	30.252
3	10	7.234	8.636	3	5.649	3.449
Total	450	450.000	449.977	450	450.000	450.000
Chi-Square		1.655	.301		1.663	0.428
D.F.		2	2		2	2
Prob.		50-30	90-80		50-30	90-80

Table 2. Simulated frequency distributions of natural resource targets for a population of N = 100; p = 0.01; n = 100.

a. Observed, expected Poisson and negative binomial frequency distributions for the original uncovered area and two simulated coverings (covered areas counted as no-target response).

Class	Expected Observed	Expected Poisson	Expected Neg. Bi.	Simulation 1 Observed	Simulation 1 Poisson	Simulation 1 Neg. Bi.	Simulation 2 Observed	Simulation 2 Poisson	Simulation 2 Neg. Bi.
0	40	36.788	39.975	51	36.788	50.642	67	36.788	67.104
1	34	36.788	33.690	30	36.788	30.235	21	36.788	20.332
2	16	18.394	16.845	11	18.394	12.572	7	18.394	7.542
3	7	6.131	6.498	6	6.131	4.468	2	6.131	2.969
4	2	1.533	2.135	2	1.899	1.453	2	1.899	1.202
>5	1	0.366	0.856	0	---	0.629	1		0.846
Total	100	100.000	99.999	100	100.000	99.999	100	100.000	99.995
X^2	---	1.564	.093	---	8.977	0.730	---	42.066	0.820
D.F.	---	3	2	---	3	2	---	3	2
Prob.	---	70-50	98-95	---	05-02	70-50	---	<0.001	70-50

Table 2. Simulated frequency distributions of natural resource targets for a population of N = 100; p = 0.01; n = 100.

b. Observed, expected Poisson and negative binomial frequency distributions for both covered and uncovered areas in two simulated coverings of the original area.

Simulation 1 18% Area Covered

	Uncovered			Covered		
Class	Observed	Poisson	Neg. Bi.	Observed	Poisson	Neg. Bi.
0	38	34.082	37.634	3	5.303	3.329
1	26	29.925	26.112	9	6.482	8.065
2	10	13.138	11.798	5	3.961	5.809
3	6	3.845	4.379	1	1.614	0.797
4	2	0.844	1.449		0.493	
>5		0.166	0.443		0.120	
>6			0.129		0.027	
7			0.036			
8			0.097			
9			0.003			
Total	82	82.000	82.000	18	18.000	18.090
Chi-Square		4.052	0.877		2.928	0.268
D.F.		2	3		2	1
Prob.		20-10	90-80		30-20	70-50

Simulation 2 54% Area Covered

	Uncovered			Covered		
Class	Observed	Poisson	Neg. Bi.	Observed	Poisson	Neg. Bi.
0	18	16.924	18.865	23	21.002	21.451
1	18	16.924	15.052	16	19.836	19.362
2	5	8.462	7.526	10	9.367	9.167
3	2	2.821	3.016	5	2.949	3.029
4	3	0.705	1.059		0.696	0.784
>5		0.164	0.340		0.132	0.169
>6			0.103		0.018	0.032
7			0.029		0.001	0.005
8			0.008			0.001
9			0.002			
Total	46	46.000	46.000	54	54.000	54.000
Chi-Square		2.170	3.186		1.371	1.000
D.F.		2	3		2	2
Prob.		50-30	50-30		70-50	70-50

Table 3. Simulated frequency distributions of natural resource targets for a population of N = 100; p = 0.05; n = 100.

a. Observed, expected Poisson and negative binomial frequency distributions for the original uncovered area and two simulated coverings (covered areas counted as no-target response).

	Original			Simulation 1			Simulation 2		
Class	Observed	Poisson	Neg. Bi.	Observed	Poisson	Neg. Bi.	Observed	Poisson	Neg. Bi.
0	0	0.674	0.533	13	0.674	6.464	44		36.223
1	1	3.369	2.966	7	3.369	12.959	1	4.043	17.544
2	9	8.422	8.063	12	8.422	16.121	6	8.422	11.453
3	19	14.037	14.282	14	14.037	15.966	11	14.037	8.120
4	16	17.547	18.533	14	17.547	13.789	10	17.547	5.985
5	13	17.547	18.784	9	17.547	10.861	8	17.547	4.512
6	18	14.622	15.479	15	14.622	8.003	7	14.622	3.452
7	14	10.444	10.661	9	10.444	5.607	6	10.444	2.669
8	4	6.528	6.261	4	6.528	3.777	4	6.528	2.079
9	3	3.627	3.183	1	3.627	2.464	1	3.627	1.630
10	3	3.183	2.253	2	3.183	3.824	2	3.183	6.136
Total	100	100.000	99.998	100	100.000	99.835	100	100.000	99.803
Chi-Square	---	8.847	8.037	---	72.911	20.886	---	433.889	38.890
D.F.	---	8	8	---	8	8	---	8	8
Prob.	---	50-30	50-30	---	<.001	.01-.001	---	<.001	<.001

Table 3. Simulated frequency distributions of natural resource targets for a population of N = 100; p = 0.05; n = 100.

b. Observed, expected Poisson and negative binomial frequency distribution for both covered and uncovered areas in two simulated coverings of the original area.

Simulation 1 34% Area Covered

	Uncovered			Covered		
Class	Observed	Poisson	Neg. Bi.	Observed	Poisson	Neg. Bi.
0	0	0.419	0.217	0	0.355	0.139
1	1	2.119	1.410	0	1.620	0.911
2	4	5.362	4.377	5	3.690	2.787
3	14	9.045	8.645	5	5.619	5.279
4	8	11.444	12.195	8	6.409	6.922
5	10	11.583	13.073	3	5.848	6.657
6	14	9.769	11.064	4	4.447	4.850
7	8	7.063	7.579	6	2.898	2.726
8	3	4.468	4.275	1	1.110	1.726
9	3	2.512	2.009			
10	1	1.271	0.954			
Total	66	65.055		32		
Chi-Square		8.427	7.739		9.527	9.385
D.F.		9	9		7	6
Prob.		50-30	70-50		30-20	20-10

Simulation 2 54% Area Covered

	Uncovered			Covered		
Class	Observed	Poisson	Neg. Bi.	Observed	Poisson	Neg. Bi.
0	0	0.324	0.344	0	0.378	0.104
1	0	1.605	1.664	1	1.875	0.834
2	5	3.978	4.043	4	4.652	3.092
3	12	6.573	6.582	7	7.697	7.007
4	6	8.145	8.078	10	9.549	10.824
5	5	8.074	7.969	8	9.479	12.035
6	6	6.670	6.585	15	7.841	9.909
7	5	4.723	4.687	6	5.559	6.116
8	3	2.926	2.933	1	3.449	2.829
9	2	1.611	1.634	2	1.902	1.251
10	2	0.799	1.471		1.370	
Total	46	45.428	45.996	54	53.751	54.00
Chi-Square		11.147	8.677		11.197	5.936
D.F.		8	8		9	7
Prob.		20-10	50-30		30-20	70-50

REGULARITY IN SPATIAL DISTRIBUTIONS: A PROBLEM OF GEOGRAPHIC ANALYSIS

MICHAEL F. DACEY
Northwestern University
Evanston, Illinois

SUMMARY

A simple stochastic location model for areal patterns is identified, and its application in geographic analysis is illustrated. The discussion of the model stresses features that characterize a class of patterns that frequently occur in geographic analysis. Because the use in geography of random counts and other stochastic models is still in its early stages of development, it is easier to identify characteristics of spatial models that remain to be studied than to cite completed works. An objective of this brief description of the nature of one class of models is to identify an area of model formulation in which geographic analysis would benefit by contributions from statisticians and others to the development, construction, and testing of appropriate models.

1. INTRODUCTION

Because the uses made by geographers of statistics, stochastic processes and other applications of probability theory are not well known outside of the social sciences, it may be useful to indicate first the role of ecological models in geographic research. Geography is both a physical science and a social science, but the social focus is becoming increasingly preeminent, and I shall stress the social science applications. Classically, the objective of geographic studies is the description of the physical and the cultural phenomena distributed over the earth's surface, and the emphasis of geographical research has been on phenomena that differentiate areas. Because the phenomena of interest are located on the earth's surface, their locations may be depicted on a map, and the map serves as the primary descriptive tool of geographic research. While geographers do many things, much of their effort is directed to the collection of data for the construction of maps and the subsequent description and analysis of mapped data. Because the map symbols indicating locations may be treated, conceptually, as points, lines or areas on a defined study area, the geographic study of regions can be formalized as the abstract study of patterns formed by points, lines and areas and of the underlying spatial processes that give rise to these patterns. While the formal study of pattern and spatial process does not dominate modern geography, an important focus of geographic research is the description of these patterns, which may extend to the study of the processes that give rise to map patterns. While a few of these processes deal with a deterministic structure and are formulated as geometric models, the study of areal distributions has tended to emphasize patterns of objects that are viewed as the realizations of stochastic processes. The description of a pattern of objects is commonly in terms of random variables that represent distance between neighboring objects (order statistics) or counts of the numbers of objects in subregions of the study region. As such, geographic research has made frequent use of tools and techniques developed by and for plant ecologists. Among the patterns geographers have studied by these methods are those formed by cities and towns within a state, the houses in rural (non-American) areas, retail store locations within cities, cultural features, public institutions such as schools, heights of land, and rivers and streams.

The use of statistical methods started in the late 1950's when the emphasis was on spacing measures for the distance between neighboring objects. In view of the setting of this meeting, it is interesting to note that the use of spacing measures in plant ecology was brought to the attention of geographers by the work of Clark and Evans [1] which was published in Science. Possibly this suggests that Science would provide an important service to the academic community by publishing more articles on mathematical and statistical techniques for analysis of specialized types of data.

The description of areal patterns based upon the counts of numbers of objects in subregions does, however, occur as early as 1932 in geographic research. Matui [6] used the method of quadrat counts to test the hypothesis that rural settlement patterns on the Tonami Plain in Japan conform to a Poisson point process. Despite this early start, it is only in the past five years that geographic research has made frequent use of random counts or the quadrat method for the study of patterns. The earlier emphasis on spacing measures seems to reflect an attitude among geographers that distance reflects an essential geographic attribute that is lacking in a frequency array showing numbers of objects in arbitrarily defined quadrats.

It should be noted that geographic applications of statistical methods have been highly derivative, and it is only very recently that geographers have attempted to develop statistical tools and stochastic models that are especially appropriate to the types of areal distributions that they study. In this paper I will briefly examine a simple model that serves to illustrate several distinctive features of geographic applications. Although the model is not new, it effectively displays the characteristics of models that are useful for the geographic analysis of a wide variety of patterns. Further, the model is useful in the sense that it fits the quadrat count data obtained from many kinds of patterns that occur in geographic analysis. First, a formulation of the model is given for a particular type of spatial pattern. Then, properties of this model are identified that characterize a wide variety of patterns of geographic interest.

2. A COUNTY SEAT MODEL FOR URBAN PATTERNS

A pattern of frequent interest to geographers is that formed by the urban places, which are treated as points, in a state or other large geographic area. The collection of urban places may be a random sample of all places listed in the census, all cities with over 10,000 population or cities defined by particular economic or social characteristics. The purpose of the model is to explain the areal pattern of a specified collection of urban places.

An important characteristic of the urban system in the United States is the dominant roles of cities that are the seat of the county government. These cities, called county seat places, are particularly important in midwestern and western United States, where large metropolitan complexes do not dominate the urban system. In midwestern and western areas, the county seat place is commonly the largest city in a county and is the focus of retail and service functions.

In most areas of the midwestern and western United States the county structure was defined before settlement. So it is not unrealistic to relate the distribution of places in a state to the county structure of that state. The simplest model assumes that all counties are identical in those factors that underlie the location of cities so that each county has equal probability of receiving each city. However, if the distinction between county seat places and other places is taken into account, it is necessary to put constraints on the occurrence of cities, because a county contains at most one place which has the county seat function. The following model, which is modified from Dacey [2], seems appropriate.

A1. Each county has an equal and independent probability of receiving a county seat place, but a county contains at most one place that is a county seat.

A2. Each county has an equal and independent probability of receiving a place that is not a county seat.

The model is further simplified by the following assumption:

A3. The probabilities that a county receives county seat places and other places are mutually independent.

The first assumption defines a sampling of counties without replacement, and the second assumption defines a sampling of counties with replacement.

Let X represent the number of places in a county, and let X_c and X_o represent the number of, repectively, county seat places and other, non-county seat places in a county. Clearly, $X = X_c + X_o$. Since a county contains at most one county seat place, put

$$P\{X_c = 1\} = p$$

$$P\{X_c = 0\} = 1 - p.$$

Locations of the two kinds of places are statistically independent, so that

$$\begin{aligned} P\{X = k\} &= P\{X_c = 1, X_o = k - 1\} + P\{X_c = 0, X_o = k\} \\ &= P\{X_c = 1\}P\{X_o = k - 1\} + P\{X_c = 0\}P\{X_0 = k\} \\ &= pP\{X_o = k - 1\} + (1 - p)P\{X_o = k\}, \end{aligned}$$

where the physical situation requires that $P\{X_o = -1\} = 0$.

Suppose X_o has the probability generating function

$$g(t) = \sum_{k=0}^{\infty} t^k P\{X_o = k\}.$$

Then the generating function of X is

$$\begin{aligned} h(t) &= \sum_{k=0}^{\infty} t^k P\{X = k\} \\ &= \sum_{k=0}^{\infty} t^k [pP\{X_o = k - 1\} + (1 - p)P\{X_o = k\}] \\ &= p \sum_{k=0}^{\infty} t^{k+1} P\{X_o = k\} + (1 - p) \sum_{k=0}^{\infty} t^k P\{X_o = k\} \\ &= ptg(t) + (1 - p)g(t). \end{aligned}$$

The moment-generating function $h(e^t)$ of X yields the moments of X, which may be expressed in terms of the moments of X_o; the mean and variance of X, obtained by standard methods, are

$$E\,X = E\,X_o + p$$

$$\text{Var } X = X_o + p(1 - p).$$

For the county seat model, A2 implies that X_o has the binomial distribution. So, if there are N counties in a state and a sample of places in the state contains M places of which $n \leq N$

are county seat places and m are non-county seat places, then $p = n/N$ and X_o has the binomial distribution with parameters m and $1/N$, or for $k = 0,1,\ldots,m+1$,

$$P\{X = k\} = \left(1 - \frac{n}{m}\right)\binom{m}{k}N^{-m}(N - 1)^{m-k} + \frac{n}{N}\binom{m}{k-1} N^{-m}(N - 1)^{m-k+1} \quad (1)$$

which has mean

$$E\ X = M/N$$

and variance

$$\text{Var } X = M/N - (n^2 + m)/N^2.$$

For many geographic applications the numbers of counties and places are large so that it is practicable to use the Poisson approximation to the binomial distribution. If $N \to \infty$, $n \to \infty$, $m \to \infty$ with $n/N = p$ and $m/N = \mu$, then

$$P\{X = k\} \to P\{Y = k\} = (1 - p)e^{-\mu}\mu^k/k! + pe^{-\mu}\mu^{k-1}/(k - 1)!. \quad (2)$$

The expected value and variance of Y are

$$E\ Y = \mu + p$$

$$\text{Var } Y = \mu + p(1 - p).$$

This model and its modifications have been tested quite extensively, and empirical analysis indicates that it describes the urban pattern of places in many agricultural or rural regions. For example, it has been tested for the M largest places in the states such as Iowa, Kansas, Mississippi, and Oklahoma for each census for which data are available. The level of correspondence between observed and calculated frequency distributions for numbers of places per county is consistently high. Table 1 compares observed and calculated frequency distributions for the M largest places in Iowa, 1950, for six values of M. The calculated frequency distribution is obtained from (2) with $p = n/N$ and $\mu = m/N$ where n, m and N are values observed for the Iowa urban system.

3. EXTENSIONS OF THE MODEL

The basic model implies (i) that counties of a state are homogeneous in those factors that underlie the density of places and (ii) that clusters of cities, such as a central city and its suburbs or satellites, do not occur. These are acceptable conditions for rural, midwestern states to which the model has been applied, but they are unacceptable for more heterogeneous or more densely populated areas.

One way to extend the model to heterogeneous regions is to allow for variation in the parameter values, p and μ of (2), by mixing or compounding of the random variable Y with different parameters. To make explicit that parameter values of (2) are to be treated as outcomes of random variables, put

$$P\{Y = k|X_1 = p_1, X_2 = t\} = (1 - p)e^{-t} \frac{t^k}{k!} + pe^{-t} \frac{t^{k-1}}{(k-1)!} .$$

The random variable X_1, for $0 \leq X_1 \leq 1$, represents the variation in the probability that the county seat place of a county is contained in a sample or other collection of urban places, and put $P\{X_1 \leq p\} = F_1(p)$. The random variable $X_2 \geq 0$ represents the variation in the density, within counties, of non-county seat places, and put $P\{X_2 \leq t\} = F_2(t)$. By taking the variation of parameter values into account, the model may be extended to urban patterns in states composed of counties that are heterogeneous in those factors that underlie the locations of urban places. One model that takes these two kinds of heterogeneity into account is defined by the following probability distribution for number of places in counties:

$$P\{Y = k\} = \int_0^1 \int_0^\infty P\{Y = k|X_1 = p, X_2 = t\}F_1\{dp\}F_2\{dt\}. \qquad (3)$$

Dacey [3] has examined the model with F_1 defined by the beta distribution and F_2 defined by the gamma distribution and has shown that this model provides a good fit to the number of urban places in a six-state, rural midwestern region (Table 2) and for a sample of United States counties that do not contain a standard metropolitan area, i.e., a city of over 50,000 population.

The extension of the model to include clusters of cities may be handled in one of two ways. If it is assumed that only county seat places are associated with clusters of cities, the extension of the model defines a random variable having the generating function

$$\psi(t) = g(t) \sum_{i=0}^{\infty} t^i \pi(i), \qquad (4)$$

where $\pi(0)$ is the probability that a county seat place of a county is not included in a sample or collection of cities, $\pi(i)$, $i \geq 1$, is the probability the sample contains a county seat place as part of a cluster of i cities, and the $\pi(i)$ are non-negative terms that sum to unity. If it is assumed that both county seat and non-seat places are associated with a cluster of cities, the extension of the model defines a random variable having the generating function

$$\psi(t) = g(tf(t)) \sum_{i=0}^{\infty} t^i \pi(i), \tag{5}$$

where $tf(t)$ is the generating function of the positive-valued random variable representing size of clusters of cities associated with non-county seat places. In both modifications of the model it is assumed that all cities forming a cluster are located in a single county, which is an acceptable assumption for states that do not contain very large metropolitan complexes.

The extensions (3), (4) and (5) of the basic county seat model to conditions of heterogeneity and urban clusters are included only to show that the county seat model may be modified in relatively simple ways to handle the patterns formed by a wide variety of urban systems. However, it is not the purpose of this paper to study the details of these models. Instead, the following paragraphs consider general characteristics of the county seat model that display several important ways that this class of models, which has relevance to geographic theory, differ from models usually constructed for plant and animal populations.

4. CHARACTERISTICS OF THE MODEL

The above discussion identifies a simple stochastic location model for the areal pattern of urban places and illustrates its application in geographic analysis. The following discussion stresses features of this model that characterize a class of patterns that frequently occur in geographic analysis.

4.1 The Space-Consuming Property.

The random variables defined by the probability distributions (1) and (2) have variance that is less than the mean. For the binomial type model, then,

$$\frac{\text{Var } X}{\text{E } X} = k - \frac{n^2 + m}{NM} < 1,$$

and for the Poisson type model

$$\frac{\text{Var } Y}{\text{E } Y} = 1 - \frac{p}{\mu + p} < 1.$$

In contrast, many of the models used in ecology have variances that are large relative to the means. The present interest, however, is on the class of patterns with relatively small variances.

These patterns appear to be formed by locations of objects that are "space consumers." They may be space consumers in the sense that the objects occupy an area that is large relative to the study area. For example, large metropolitan centers occupy a substantial part of county area. They may also be space consumers in the sense that the existence of the objects depends upon exclusive domination of a surrounding space. For example, the economic support of most small, midwestern central places was until recent years the provision of retail and service functions to the rural population in its surrounding hinterland. The maintenance of a level of economic support for a viable central place sets a lower bound on hinterland area, which depends on the density of the rural population. This threshold size, along with the agglomeration and scale economies associated with retail and service functions, effectively precludes the establishment of a new central place in close proximity to an already established central place. So, until suburban complexes begin to develop around a large city, the successful founding of a retail and service oriented town concomitantly delimits a larger surrounding area in which it is unlikely that another central place will be able to survive. A county seat place is clearly of this central place type because it is the center of a wide variety of governmental activities that are provided for the entire county.

Locations of objects that are space consumers tend toward an even or regular distribution. In turn, the frequency distribution of quadrat counts obtained from this type of pattern tend to have a small variance, and this type of pattern is sometimes called, in the ecologic literature, one of over-dispersion. For the phenomena of interest to ecologists, it appears that over-dispersion occurs only when a population has occupied a homogeneous area for a long time. In contrast, patterns of urban places in midwestern United States are over-dispersed from the time when each state was opened to settlement and this over-dispersion continues for many decades. When clustering of cities does occur, it occurs only as a late stage of settlement associated with the declining importance of rural populations to the economic support of cities. This sequence of colonization and growth is displayed in the historic series of quadrat count data given by Hudson [5] for settlement patterns in Iowa for the 1870-1969 period.

While the above discussion is restricted to the space consuming properties of urban places in agricultural regions, many patterns studied by geographers are composed of objects that tend to generate an areal domain in which the occurrence of other objects, of the same kind, is restricted or precluded. For example, one of the patterns studied in the early article by Matui [6] is formed by farms. Farms are space consumers and the observed frequency distribution of isolated farms, Table 3, among subregions formed by partitioning the study area into small quadrats corresponds closely to the frequency distribution calculated from (2).

4.2 The Natural Partioning of a Region.

A difficulty in applying random count models to the areal patterns formed by a collection of objects is the selection of an appropriate size for the subregions in which objects are counted. There are problems that arise at the theoretical level in the construction of a model and at the empirical level in the specification of a sample design. Many of the patterns of interest to geographers are imposed on a study region that has a "natural" partition into subregions, and these subregions form the areal base for the construction and analysis of random count models. These subregions provide a fundamental areal unit on which to base the construction of random count data. More importantly, these subregions provide an areal basis for the construction of theoretical models since the availability of this partition facilitates the physical interpretation of random count models that generate discrete probability laws.

In the case of the county seat model, the division of counties into states is a "natural" partition in the sense that midwestern states were partitioned into counties before they were open to settlement. Because the location of county boundaries preceded the founding of towns it is not unreasonable to examine the validity of a model that relates the urban pattern to the county structure. The availability of a natural partition also facilitates the interpretation of the implications to geography theory of the finding that the county seat model has close correspondence with empirical evidence.

In addition to the counties of a state, these natural partitions available to the analysis of the patterns of cultural phenomena include subregions such as the townships which typically partition a county and the plotting of townships into blocks. Other partitions may be based on physical characteristics of regions such as soil, climate or local relief. The important fact is that partitions of these types exist prior to settlement and the subsequent location of cultural phenomena. When these partitions exert influence on the locations and density of objects, they provide the natural areal units to be incorporated in random count models of location processes.

4.3 Correlation Effects.

When random count models are based on arbitrarily defined subregions, it may be difficult to invoke theoretical considerations in specifying the contiguity or correlation effects between the numbers of objects in adjacent subregions. The availability of a natural partition may facilitate, at one level, the theoretical specification of the degree and nature of correlation effects and, at another level, the empirical measurement of, and testing for, correlation effects. The county seat model implies that the number of objects in a county is independent of the numbers of objects in either adjacent or distant counties. The contiguity measure proposed by Moran [7] for the statistical interpretation of maps was used to evaluate the validity of this assumption for the Iowa urban pattern, Dacey [4], and extensive tests presented no evidence contrary to the assumption. This contiguity test, along with the comparison of the observed and calculated frequency arrays, permits more stringent evaluation of hypotheses than does only the use of quadrat counts. Because their training emphasizes the description and analysis of areal relations, geographers will tend to be skeptical of theoretical models or empirical analyses that do not take correlation effects into account. In this regard, it is significant that the analysis of pattern in Matui's early use of random count models is largely concerned with the use of runs and other tests as a means for testing the presence of correlation effects in the areal arrangement of quadrat counts.

4.4 The Sample Population.

A final characteristic, it must be admitted, is that the patterns of geographic interest frequently comprise a census that enumerates all of the specified objects within a study area. Despite this fact, statistical methods of large sample theory are often used to evaluate hypotheses. While it is assumed there exists a population from which a sample is drawn, often it is embarrassingly difficult to identify this population.

The failure to recognize the underlying population in part explains the absence in Tables 1, 2, and 3 of statistical tests of goodness-of-fit.

ACKNOWLEDGMENT

The support of the National Science Foundation, Grant GS-1627, is gratefully acknowledged.

REFERENCES

[1] Clark, P. J., and Evans, F. 1952. On some aspects of spatial patterns in biological populations. Science 121:397-8.

[2] Dacey, M. F. 1964. Modified Poisson probability law for point pattern more regular than random. Annals A.A.G. 54, 559-565.

[3] ———. 1966a. A compound probability law for a pattern more dispersed than random and with areal inhomogeneity. Economic Geog. 42:172-9.

[4] ———. 1966b. A county-seat model for the areal pattern of an urban system. Geog. Rev. 56:527-42.

[5] Hudson, J. C. 1969. A location theory for rural settlement. Annals A.A.G. 59:365-81.

[6] Matui, I. 1932. Statistical study of the distribution of scattered villages in two regions of the Tonami Plain, Toyama Prefecture. Japanese Jour. Geol. Geog. 9:251-66.

[7] Moran, P. A. P. 1948. The interpretation of statistical maps. Jour. Roy. Statist. Soc., Ser. B, 10:243-51.

Table 1. Frequency Distributions of Places among Counties for Places in Iowa, 1950.

Parameters of Equation (1)						
M	20	40	60	80	100	200
n	18	34	50	66	78	98
m	2	6	10	14	22	101
N	99	99	99	99	99	99

Places per County	Number of Counties											
	ob	cal	ob	cal	ob	cal	ob	cal	ob	cal	ob	cal
0	80	80	60	61	43	44	28	29	15	17	0	0
1	18	19	38	36	52	50	62	61	69	66	34	35
2	1	0	1	2	4	5	9	8	14	14	38	36
3								1	1	2	21	19
4											3	6
5											3	2
6											0	1

Notes:

ob - observed frequency distribution

cal - frequency distribution calculated from equation (1)

These data are adapted from Dacey [4].

Table 2. Frequency Distribution of Places among Counties in a Six-State Midwestern Region

Places per County	Observed	Calculated
0	211	222
1	222	217
2	40	47
3	12	9
4	2	2
$\geq$ 5	0	0

Notes:

The calculated frequency distribution was obtained from (3) with parameters estimated by the method of moments.

These data are adapted from Dacey [3].

Table 3. Frequency Distributions of Isolated Farms in Quadrats of the Tonami Plain

Farms per Quadrat	Observed	Calculated
0	206	205
1	318	318
2	164	165
3	49	50
4	12	10
5	1	2
≥ 6	0	0

Notes:

Observed data are from Matui [6].

Calculated values are from (2) with parameters estimated by method of moments.

SOME STATISTICAL ASPECTS OF AMOUNTS AND DURATION OF RAINFALL*

L. R. SHENTON
P. SKEES
Computer Center
University of Georgia

SUMMARY

Measurements associated with meterological phenomena are receiving increasing attention because of the importance of weather modification experiments. In particular, attempts have been made to suppress the intensity of hail storms, to diffuse and disperse borderline and below visibility conditions, and to increase or decrease precipitation whenever conditions are critically poised for its formation.

A question of considerable importance concerns the statistical evaluation of differences of rainfall amounts between two locations (stations) or at two groups of stations (control area and target area) under controlled experimental conditions. This paper refers to rainfall for a number of stations in the Southeastern United States (mostly in Georgia) and mainly concerns: (a) transformations of amounts per storm which produce near normal distributions, or distributions with reduced skewness and kurtosis. (b) distributions of the duration (in hours and days) of rain storms.

*This work was partially supported through the Southeastern Forest Experiment Station with Department of Defense Funds, IPR19-8-8008-WEPS.

1. INTRODUCTION

Measurements associated with meterological phenomena are receiving increasing attention because of the importance of weather modification experiments. In particular, attempts have been made to suppress the intensity of hail storms, to diffuse and disperse borderline and below visibility, and to increase or decrease precipitation whenever conditions are critically poised for its formation. Evidence of statistical activity is displayed in the recent Berkeley Symposium Volume [2], and a survey of statistical aspects of cloud seeding is to be found in Neyman [7] and the related references.

A question of considerable importance concerns the statistical evaluation of differences of rainfall amounts between two locations (stations) or at two groups of stations (control area and target area) under controlled experimental conditions. There is an extensive literature on this subject, but an underlying problem which seems to be of prime importance concerns the actual distribution of rainfall. This can be studied from the point of view of (a) amounts per unit of time, (b) amounts per storm, and (c) duration. The unit of time is to a large extent a function of instrumentation and data processing; many stations record daily amounts, a considerable number record hourly amounts, and a few are equipped to record over intervals as small as ten minutes (the smallest amount recorded is usually 0.01 inches.) It is evident that duration of rainfall is not a simple concept because (especially for hourly or smaller records) of its discontinuous nature; thus a storm of two days duration could arise even when the second day only contributed a trace. When hourly amounts are available we can define a storm as a period of rain separated by 1, 2,..., 24 hour(s) of dryness. For example Changnon and Huff [2] use a six hour definition, so a 'storm' is a rain period separated from preceding and suceeding rainfall by six hours or more.

Much work on the distribution of rainfall amounts has been carried out by Thom ([13 - 15] and related references), who has put forward reasons for using the 2-parameter gamma distribution. He considers, among other things, rainfall amounts per storm, and also rainfall amounts per month at 122 first-order weather stations over a 30-year period (Thom, [15 - 17]), introducing approximate maximum likelihood estimators which turn out to have surprisingly good properties (Shenton and Bowman, [12]). Thom also uses the gamma distribution with zeros added for the distribution of rainfall amounts when dry days (days with zero rainfall) are included.

A general account of the usefulness of several statistical distributions, as applied to annual rainfall amounts, is to be found in Markovic [6].

Our study refers to rainfall for a number of stations in the Southeastern United States (mostly in Georgia) and mainly concerns:

a) transformations of amounts per storm which produce near normal distributions, or distributions with reduced skewness and kurtosis.

b) distributions of the duration (in hours and days) of rain storms.

2. BEST TRANSFORMATIONS OF STORM AMOUNTS

A set of 70 or 80 transformations have been applied to storm amounts (hourly observations) for three stations, the object being to reduce the skewness and kurtosis to zero and three, respectively, (normal values) or to produce a good chi-square approximation (we use, as usual, $\sqrt{\beta_1} = \mu_3/\mu_2^{3/2}$ for the skewness, and $\beta_2 = \mu_4/\mu_2^2$ for the kurtosis, where μ_2, μ_3, μ_4 are the first three central moments). The transformations belonged to the following sets:

(a) $\{\ln(a + bx)\}^m$; $a \geq 0$, $b > 0$, $m = 0.1(0.1)\ 1(1)\ 4$

(b) $(a + bx)^m$, $m = 0.1(0.1)\ 1.0$, $a > 0$, $b > 0$

(c) $x^m - (x_{max} - x)^m$; $m - 0.1(0.1)\ 2.0$

(d) sinh x, cosh x, arctan x, cosech x, tanh x.

It has previously been assumed that the cube-root of rainfall (Stidd) is nearly normally distributed, but we did not find this to be the case; the result for a number of Georgia stations over periods of 15 or more years and for the cube-root of daily amounts of rainfall with and without dry days, are graphically displayed in Figures 1 and 2; the plots refer to the value of skewness (β_1) and kurtosis (β_2) and show marked departures from normality ($\beta_1 = 0$, $\beta_2 = 3$).

Some of the results for the transformed rainfall amounts per storm (with 1-hour and 8-hour storm definitions) are shown graphically in Figure 3, using $\sqrt{\beta_1}$, β_2 as axes. Evidently it is possible to produce a distribution near enough to Pearson's Type III (the χ^2 or gamma distribution), but the transformation is usually of class (c), and therefore not simple to apply. Other favorable transformations are given in Table 3, and perhaps the simplest to apply is lnx, which for the class considered produces a Pearson Type I curve (equation of the form $y = (X - \alpha)^a(\beta - X)^b$). The best "normalizing" transformations seem to vary considerably from station to station and also with respect to storm definition.

As examples of the transformed distributions we quote the following:

<u>Greensboro</u> Transformation $t = x^{0.1}$ storm def. = 1 hr.
Pearson distribution $y = c(\sqrt[10]{x} - 0.6030)^{0.1801}$ $(1.0937 - \sqrt[10]{x})^{0.5357}$

(Limits of x are 0.006 to 2.45 approx.)

<u>Charleston</u> Transformation $t = \ln x$ storm def. = 1 hr.
Pearson distribution $v = c(5.2564 + \ln x)^{0.6053}$ $(1.3968 - \ln x)^{0.9130}$

(Limits of x are 0.01 to 4.00 approx.)

If we used the transformed variates with the appropriate Pearson curve, then approximate confidence limits could be set up from the tables of Johnson, et al. [3]. We find for example the following:

<u>Greensboro</u> storm def. = 1 hr.

95% limits for rainfall per storm

Transformation	Limits
Untransformed	$-0.62 < x < 1.08$
ln x	$0.007 < x < 1.363$
x 0.1	$0.006 < x < 1.465$

<u>Charleston</u> storm def. = 1 hr.

Transformation	Upper 5% limit on x	Upper 1% limit on x
Untransformed	0.812	1.544
x 0.1	1.497	2.792
ln x	1.530	2.647

These suggest that the application of the usual normal theory to fixing limits on rainfall amounts per storm <u>yields unduly optimistic low values for upper limits</u>, and these may be in error by as much as 75% to 100%. An interesting application of confidence intervals to rainfall amounts is to be found in Manning [4, 5]; see also Thom [15].

3. DURATION OF STORMS

3.1 Appropriate Distributions.

Since the duration of storms (using daily observations, or hourly with different storm definitions) usually follow a J-shaped distribution, and since the least period involved is a day (or hour), we need only consider those discrete distributions with no zero class and with few (zero or a half) modes. This limits the class of suitable distributions to a few possibilities, especially if we also wish to confine attention to distributions with simple structures. An obvious choice is the logarithmic series distribution (L.S.D.) and this has been applied to rainfall distributions with some success by Patil ([9] and related references). The distributions and their parametric forms used were as follows (P_x being used for the probability function):

(a) Logarithmic Series Distribution (L.S.D.)

$$P_x = \frac{\theta^x}{xk} \qquad 0 < \theta < 1$$

$$\text{Mean} = \theta/(1-\theta)k \qquad k = -\ln(1-\theta)$$

(b) Negative Binomial Distribution (N.B.D.)

$$P_x = \binom{k+x-1}{x}\left(\frac{p^x}{(1+p)^{x+h}}\right) \qquad (k > 0,\ p > 0)$$

$$\text{Mean} = 1 + kp$$

(c) Geometric Distribution (G.D.)

$$P_x = (1-p)p^{x-1} \qquad (0 < p < 1)$$

$$\text{Mean} = 1/(1-p)$$

(d) Geometric Gram-Charlier Distribution (G. G-C. D.)

$$P_x = (1-p)p^{x-1}\ \{1 + \lambda(x - \frac{1}{1-p})\} \quad (0 < \lambda < \frac{1}{p} - 1)$$

$$\text{Mean} = \frac{1}{q} + \frac{\lambda p}{q^2}$$

Probability Generating Function $[G(t) = E\ t^x]$

$$G(t) = \frac{qt}{1-pt} + \frac{\lambda pt(t-1)}{(1-pt)^2}$$

(e) Geometric - Logarithmic Mixture (G.L.M.)

$$P_x = a(1-p)p^{x-1} + b\theta^x/x \qquad (0 \le p < 1,\ 0 \le \theta < 1)$$

$$\text{Mean} = \frac{a}{q} + \frac{b\theta}{1-\theta}$$

(f) Logarithmic Negative Mixture (L.N.M.)

Distribution Function

$$\sum_{y=1}^{x} P_y = 1 - \frac{a\theta^x}{x+b} \qquad (P_1 = 1 - \frac{a\theta}{b+1})$$

$$P_x = a\theta^{x-1}\left\{\frac{1}{x+b-1} - \frac{\theta}{x+b}\right\} (x = 2,3,\ldots)$$

$$(a > 0,\ b > -1,\ 0 < \theta < 1)$$

$$\text{Mean} = 1 + a \sum_{s=1}^{\infty} \frac{\theta^s}{b+s}$$

(In each of the above distributions, the range of the variate is $x = 1,2,\ldots,\infty$)

3.2 Methods of Estimation. For (a) it is simplest to use the mean (which is the maximum likelihood estimator) although an unbiased minimum variance estimator (Patil, [8, 9]; Patil, Kamat and Wani, [10]) does exist (some additional comments on distributional properties of the maximum likelihood estimator are given by Shenton and Bowman in Volume 1 of this trilogy). For (c) the mean is the estimator, whereas for (b) and (d) the mean and variance or mean and first frequency can be used. The geometric-logarithmic mixture is more difficult to deal with. Avoiding higher moments as estimators we can use the first two relative frequencies n_1^*, n_2^*, and the mean m_1' as follows:

$$f_0 = a - b\ln(1-\theta) = 0$$

$$f_1 = aq + b\theta - n_1^* = 0$$

$$f_2 = aq(1-q) + b\theta^2/2 - n_2^* = 0$$

$$f_3 = a/q + b\theta/(1-\theta) - m_1'$$

Solutions can now be found by eliminating a,b,q leading to the equations

$$aP - bPQ + cQ^2 = 0,$$

or $$aP - BPQ + CQ^2 = 0,$$

where

$$a = m_1'L + \theta/(1-\theta), \qquad (L = \ln(1-\theta))$$

$$b = m_1'\theta - n_1^*\theta/(1-\theta),$$

$$c = -(\theta + n_1^*L),$$

$$A = -c,$$

$$B = L(n_2^* - n_1^*) - \theta(1-\theta/2),$$

$$C = n_2^*\theta - n_1^*\theta^2/2,$$

$$P = aC - Ac, \ Q - aB - Ab.$$

Another method of solution is to use a digital search technique (see for example Saaty and Bram, [11]) to find the minimum of

$$S = f_0^2 + f_1^2 + f_2^2 + f_3^2 .$$

The logarithmic negative mixture (f) is suggested from Burr's [1] approach to frequency distributions. The three parameters a,θ,b may be estimated by using three frequencies, and choosing those which minimize (using the work in a broad sense) chi-squared. Of course it would be possible to use maximum likelihood procedures or moments, but the frequency approach seems simplest. With this in mind let n_r^*, n_s^*, n_t^*, be three relative frequencies ($r \neq s \neq t$), and write $N_r^* = 1 - \sum_1^r n_m^*$, etc. Then the parameters are estimated from

$$\left(\frac{N_r^*}{N_s^*} \cdot \frac{b+r}{b+s}\right)^{\frac{1}{r-s}} = \left(\frac{N_r^*}{N_t^*} \cdot \frac{b+r}{b+t}\right)^{\frac{1}{r-t}}$$

$$\theta = \left(\frac{N_r^*}{N_s^*} \cdot \frac{b+r}{b+s}\right)^{\frac{1}{r-s}}$$

$$a = \frac{(b+r)\ N_r^*}{\theta^r}$$

In passing, we remark that all higher moments of the logarithmic negative mixture may be expressed in terms of the mean μ_1. For the probability generating function is

$$G(t) = Ft^x$$

$$= t + a(t-1) \sum_{s=1}^{\infty} \frac{\theta^s t^s}{b+s} ,$$

so the rth factorial moment is

$$\mu_{(r)} = r\ a \sum_{s=1}^{\infty} \frac{s(s-1)\ \ldots\ (s-r+2)\theta^s}{b+s}, \ (r = 2,3,\ldots),$$

from which

$$\mu_{(2)} = 2a \left(\frac{\theta}{1-\theta} + b(1-\mu_1)\right) .$$

By differentiating G(t) with respect to t, replacing t by $1+\alpha$ and selecting the coefficient of α^s, we have for the higher factorial moments

$$\nu_{(r+1)} = -\ (r+b-1)\nu_{(r)} + \frac{a(r-1)!}{\theta} \left(\frac{\theta}{1-\theta}\right)^r, \quad (r = 2,3,\ldots)$$

where

$$\mu_{(r)} = r\ \nu_{(r)} .$$

3.3 Storms Measured in Units of a Day

In this case we need not be too concerned with the storm definition, for 24 hours without rain (measured from midnight, for example) would appear to be acceptable as an indication of storm termination. We have fitted the logarithmic distribution by maximum likelihood, and the negative binomial distribution by moments, to the data for sixteen reporting stations in the Southeastern States (mostly in Georgia). A summary of the χ^2 goodness-of-fit analysis was as follows:

Acceptance Level	Logarithmic Series Dist.	Neg. Binomial Dist.
5%	3	10
1%	2	2
Rejected	11	4

Evidently the logarithmic distribution is not generally applicable, where as the negative binomial distribution seems better suited to the structure. We feel that the daily unit for recording storm duration is perhaps too long and more attention should be given to 'hourly' storm distributions.

3.4 Storms Measured in Units of an Hour

In this case three hourly recording stations Athens, (Ga.), Charleston (S.C.) and Greensboro (N.C.) were analyzed for eight storm definitions (Tables 4a, 4b, 4c). Taking the upper 5% value of χ^2 as the acceptance level, the results for the 24 distributions came out to be: no acceptance for the G.D. and G.G.-C.D., one acceptance for the L.S.D., one acceptance for the N.B.D. and twenty-two acceptances for the L.N.M. In addition a few cases were tried with a mixture of geometric distributions and in this case there were 3 "successes" out of 8. Similarly, trials with the G.L.M. did not give encouraging results.

Since the L.N.M. was by far the most successful distribution investigated, we give in Table 5 the estimated parameters (b, θ, a) and the cumulated frequencies ($N_s = \sum_{r=1}^{s} n_r$) employed for the twenty-four sets of data. In each case about 15-25 frequency combinations were tested from the point of view of minimizing χ^2, including obvious cases such as N_1, N_2, N_s with $s = 3(1)15$; combinations which ignored n_1 were rarely successful. It will be noticed that the parameter b (a displacement factor which would be zero for the L.S.D. itself) tends to increase in step with the number of hours in the storm definition, and that the value of θ is quite near to unity. Of course θ near to unity corresponds to a long-tailed distribution, whereas b acts as a damping factor.

The computed distributions for 8 of the 24 cases are given in Table 6.

4. DISCUSSION

The records we have studied of rainfall show that amounts per day or per hour or per storm follow sharply skewed distributions. Moreover, in general, the standard deviation is about equal to or even slightly larger than the mean itself. This implies that significance tests based on normal theory are not applicable, even approximately, and use should be made of best fitting Pearson distributions along with the tables of their percentage points (Johnson, et al. [3]). Transformations of the form ln x or $x^{0.1}$ significantly reduce the skewness and kurtosis and tend to produce Beta-type distributions with more or less stable characteristics; whether this applies to other locations is an open question. In this respect we point out that the tenth-root of rainfall per hour for three widely separated locations has s.d. close to 0.13 inches.

It is quite difficult to find a good normalizing transformation for storms based on hourly data, skewness being a very persistent characteristic. Moreover it seems unlikely that the same transformations (such as $x^{0.1}$, ln x, etc.) which by and large dilute skewness and kurtosis.

As for the duration of storms (hourly observations being available) none of the classical discrete distributions seem suitable. The reason for this could relate to the J-shaped distributions invariably encountered; very rarely one finds a distribution with a mode far from the start. This feature rules out the negative binomial, the Neyman-Type A and many similar compound distributions. The logarithmic series distribution, depending as it does on one parameter, lacks flexibility. We have introduced a modified logarithmic distribution, admittedly with little physical justification, with cumulative distribution of the form $F(x) = 1 - a\theta^{x}/(x+b)$; and this leads to a mixture of logarithmic-type distributions with negative mixing. It turns out that most of the duration data (hours being the units) is fitted rather well by this modified logarithmic distribution.

5. ACKNOWLEDGMENTS

We feel indebted to the Meterological Staff at the Macon Fire Laboratory and in particular to Jim Paul for much stimulating discussion. Of course, the views expressed in this paper are entirely our responsibility.

REFERENCES

[1] Burr, I. W. 1942. Cumulative frequency functions. Ann. Math. Statist., 13:215.

[2] Changnon, J., and Huff, F. A. 1967. The effect of natural rainfall variability in verification of rain modification experiments. Fifth Berkeley Symposium on Mathematical Statistics and Probability: Volume V, Weather Modification, 177-98.

[3] Johnson, N. S., Nixon, E., Amos, D. E., and Pearson, E. S. 1963. Table of percentage points of Pearson curves for given $\sqrt{\beta_1}$ and β_2 expressed in standard measure. Biometrika, 50:459.

[4] Manning, H. L. 1951. Confidence Limits of Expected Monthly Rainfall. Research Memoir, Empire Cotton Growing Corporation, London.

[5] Manning, H. L. 1956. The Statistical Assessment of rainfall probability and its application in Uganda agriculture. Proc. Roy. Soc. (London) B, 144, 460-79.

[6] Markovic, Radmilo D. 1965. Probability Functions of best fit to Distributions of Annual Precipitation and Runoff. Hydrology Papers, Colorado State University, Fort Collins, Colorado.

[7] Neyman, J. 1967. Experimentation with weather control (with Discussion). J.R.S.S., A, 130:285-326.

[8] Patil, G. P. 1961. Some methods of estimation for the logarithmic distribution. Biometrics, 18:68-75.

[9] ———. 1964. On minimum variance unbiased estimation for the logarithmic series. Bull. Inst. Internat. Statist., 40:766.

[10] ———; Kamat, A. R.; and Wani, J. K. 1966. Certain studies on the structure and statistics of the logarithmic series distribution and related tables. Tech. Rep. Aerospace Res. Lab., Wright-Patterson Air Force Base, p. 389.

[11] Saaty, T. L., and Bram, J. 1964. Nonlinear Mathematics, Chapter 2. McGraw-Hill Book Company, New York.

[12] Shenton, L. R., and Bowman, K. O. 1969. Remarks on Thom's Estimators for the Gamma Distribution. Monthly Weather Review. (To appear).

[13] Thom, H. C. S. 1957. A Statistical Method of Evaluating Augmentation of Precipitation by Cloud Seeding, Final Report of the Advisory Committee on Weather Control, Volume 2.

[14] ———. 1958. A Note on the Gamma Distribution. Monthly Weather Review, Vol. 86, No. 4, pp. 117-22.

[15] ———. 1968. Approximate Convolution of the Gamma and Mixed Gamma Distributions. Monthly Weather Review, Vol. 96, No. 12, pp. 883-6.

[16] ———. 1968, April. Direct and Inverse Tables of the Gamma Distribution. ESSA Tech. Report, EDS-2.

[17] ———. 1968, August. Quantiles of Monthly Precipitation for Selected Stations in the Contiguous United States. ESSA Tech. Report, EDS-6.

Table 1. Skewness and Kurtosis for Amounts of Daily Rainfall* under several Transformations. Georgia Stations

$\sqrt{\beta_1} = \mu_3/\mu_2^{1.5}$, $\beta_2 = \mu_4/\mu_2^2$ (μ_s is s^{th} central moment)

Station (coded)	$\sqrt{\beta_1}$	β_2	$2\beta_2-3\beta_1-6$	Trans.	$\sqrt{\beta_1}$	β_2	Number of days
0131	2.78	16.3	3.4	$\sqrt[3]{x}$	0.34	3.01	6472
0140	3.13	19.2	3.0	$\sqrt[5]{x}$	-0.76	3.12	7500
0253	3.93	34.1	15.8	$\sqrt[5]{x}$	-0.76	3.05	4000
0430	3.13	21.1	6.8	$\sqrt[4]{x}$	-0.21	2.96	4000
1982	2.64	13.3	-0.3	$\sqrt[4]{x}$	-0.22	3.18	1700
2166	3.35	19.7	-0.3	$\sqrt[3]{x}$	0.31	2.31	2000
2839	4.47	51.4	29.4	$\sqrt[4]{x}$	-0.20	3.34	3300
2966	2.50	12.0	-.75	$\sqrt[3]{x}$	0.19	2.82	1062
2475	3.38	20.6	0.9	$\sqrt[5]{x}$	-0.60	2.75	3723

(*Dry days are not included in the samples)

Table 2. Rainfall Amounts for Storms (Hourly Data, 1955-64) Moments for Several Southern Stations

Station	Storm def.	Mean $\bar{x}$	S.D. σ	Skewness $\sqrt{\beta_1}$	Kurtosis β_2	Type III Criterion $2\beta_2-3\beta_1-6$
Athens, Ga.	1	0.3358	0.5918	3.71	21.65	-3.97
	2	0.3955	0.6452	3.37	18.24	-3.58
	6	0.5249	0.7690	3.17	17.28	-1.58
	8	0.5697	0.8023	2.97	15.44	-1.65
Charleston S.C.	1	0.3328	0.5988	4.13	27.49	-2.15
	2	0.3868	0.6527	3.79	23.34	-2.53
	3	0.4259	0.6861	3.56	20.84	-2.33
	4	0.4526	0.7111	3.48	19.83	-2.60
	5	0.4769	0.7386	3.45	19.41	-2.81
	6	0.4957	0.7557	3.32	18.22	-2.69
	7	0.5165	0.7716	3.22	17.26	-2.62
	8	0.5357	0.7807	3.15	16.71	-2.39
Greensboro N.C.	1	0.2306	0.4332	4.60	35.77	1.93
	2	0.2749	0.4769	4.14	29.42	1.49
	3	0.3066	0.5010	3.91	26.84	1.73
	4	0.3308	0.5226	3.73	24.33	1.01
	5	0.3524	0.5486	3.69	23.22	-0.44
	6	0.3687	0.5634	3.50	21.26	-0.27
	7	0.3850	0.5733	3.42	20.42	-0.26
	8	0.3973	0.5829	3.40	20.37	0.05
Knoxville Tenn.	1	0.4732	0.4005	3.59	18.03	-8.63
	2	0.5420	0.4110	3.63	16.87	-11.83
	3	0.5899	0.4206	3.62	15.50	-14.37
	4	0.6202	0.4248	3.67	14.90	-16.71
	5	0.6488	0.4560	4.22	21.13	-17.12
	6	0.6764	0.4519	4.45	21.78	-21.91
	7	0.7086	0.4621	4.69	22.58	-26.91
	8	0.7284	0.4755	4.41	20.80	-22.97
Valdosta, Ga.	1	0.3704	0.6161	3.59	20.72	-3.18
	2	0.4227	0.6647	3.45	19.49	-2.74
	6	0.5150	0.7502	3.47	19.07	-1.47
	8	0.5361	0.7667	3.23	17.95	-1.42

(Storm definition: x denotes the number of hours dryness between successive storms (periods of measurable rain).)

Table 3. Rainfall Amounts for Storms (Hourly Data: 1955-64)
Some Skewness and Kurtosis reducing transformations

Station	S.D.	Transformation	Moments of transformed variate μ_1'	σ	$\sqrt{\beta_1}$	β_2
Athens, Ga.	1	$\sqrt[10]{x}$	0.8062	0.1328	0.41	2.20
		$\ln x$	-2.2879	1.6259	0.18	1.95
		$\sqrt[4]{(1+x)^3} - \sqrt[4]{x^3}$	0.8614	0.1126	-0.97	3.04
Charleston S.C.	1	$\sqrt[10]{x}$	0.8107	0.1278	0.40	2.38
		$\sqrt[5]{\ln(1+x)}$	0.6517	0.1838	0.37	2.19
		$\ln(1+x)$	0.2294	0.3025	2.18	8.38
		$\ln x$	-2.2207	1.5599	0.14	2.10
		$\sqrt[5]{x}$	0.6736	0.2145	0.69	2.89*
	6	$\sqrt[10]{x}$	0.8552	0.1297	0.09	2.29
		$\ln x$	-1.6807	1.5386	-0.19	2.24
	8	$\sqrt[10]{x}$	0.8657	0.1285	0.03	2.32
Greensboro N.C.	1	$\sqrt[10]{x}$	0.7738	0.1276	0.54	2.23
		$\ln x$	-2.6964	1.6091	0.33	1.90
		$\sqrt[2/5]{\ln(1+x)}$	0.3932	0.2407	0.92	2.98*
	8	$\sqrt[10]{x}$	0.8362	0.1292	0.04	2.09
		$\ln x$	1.9110	1.5714	-0.20	2.03
Valdosta Ga.	1	$\sqrt[10]{x}$	0.8267	0.1244	0.35	2.38
		$\ln x$	-2.0158	1.4959	0.08	2.19
		$\sqrt{\ln(1+x)}$	0.4236	0.2734	0.93	3.18*
	6	$\sqrt[10]{x}$	0.8668	0.1222	0.09	2.41
		$\ln x$	-1.5305	1.4280	-0.20	2.42

(S.D. ≡ Storm definition. Entries marked with * are those transformations out of 70 or more which came nearest to the normal point $\beta_1 = 0$, $\beta_2 = 3.0$)

Table 4a. Distribution of Storm Duration (Hourly data: Athens, G., 1955-64) Storm Definition (x dry hours between storms)

Dur. (hrs.) x	1	2	3	4	5	6	7	8
1	363	248	196	163	139	129	120	110
2	291	225	192	170	148	140	130	125
3	142	133	119	114	102	100	92	86
4	71	79	79	67	71	69	64	67
5	51	51	50	50	54	52	49	42
6	38	36	31	32	32	27	26	25
7	32	25	28	29	29	27	25	25
8	25	29	24	25	24	24	27	28
9	16	26	22	24	24	22	22	22
10	16	11	14	19	19	19	19	17
11	12	18	19	19	15	15	12	12
12	14	12	11	13	15	15	18	20
13	10	11	9	9	9	9	10	4
14	8	4	6	6	8	4	5	5
15	3	6	8	8	7	8	8	10
16	4	8	8	9	10	11	8	9
17	1	2	4	4	2	5	7	6
18	3	6	7	8	8	8	8	10
19	0	1	1	1	5	2	2	3
20	3	1	1	0	0	1	1	3
21	3	3	6	6	7	9	9	8
22	1	2	3	3	3	4	3	3
23	2	4	3	1	1	2	2	2
24	1	0	3	3	2	2	3	2
25	0	0	1	1	1	1	0	2
26	1	1	1	1	2	3	2	2
27	1	2	2	1	1	0	2	1
28	0	0	0	0	0	1	2	2
29	0	0	0	1	1	1	1	1
30	1	0	0	0	0	0	0	0
31	0	0	0	0	0	0	0	1
32	0	1	1	1	0	0	0	0
33	0	0	0	0	0	0	0	0
34	0	0	0	0	1	0	0	0
35	0	0	0	1	1	1	2	2
36	0	0	0	0	0	0	0	0
37	0	0	0	0	0	0	0	0
38	0	0	0	0	0	0	0	0
39	0	0	0	0	0	0	0	0
40	0	0	0	0	0	0	0	0
41	0	0	0	0	0	0	0	0
42	0	0	0	0	0	0	0	0
43	0	0	0	0	0	1	1	1
$\bar{X}$	3.4915	4.1122	4.5826	4.9252	5.2443	5.4579	5.7147	5.9238
σ	3.7573	4.2764	4.7673	4.9873	5.1992	5.5357	5.8290	6.0327

Table 4b. Distribution of Storm Duration (Hourly data: Charleston, N.C., 1955-64) Storm Definition (x dry hours between storms)

	1	2	3	4	5	6	7	8
1	538	407	328	292	258	233	208	187
2	419	365	304	266	246	238	227	210
3	274	235	217	198	193	191	182	174
4	157	148	149	144	132	118	109	109
5	87	87	89	91	87	81	80	83
6	49	60	62	59	64	65	58	57
7	40	35	40	43	42	42	49	52
8	27	36	37	43	41	43	43	40
9	21	19	17	24	24	23	27	30
10	16	17	21	15	18	16	20	18
11	13	16	16	20	15	19	14	14
12	7	12	8	10	12	13	13	13
13	8	7	9	10	13	14	16	15
14	2	5	7	8	8	8	11	11
15	11	14	14	12	13	11	12	11
16	5	9	11	13	14	15	10	12
17	2	2	4	4	3	4	7	10
18	2	2	4	2	2	2	2	2
19	1	4	5	6	7	6	7	6
20	1	1	1	3	3	5	5	4
21	2	0	1	0	2	4	2	1
22	3	5	4	4	4	4	4	6
23	2	2	0	1	1	0	0	0
24	1	2	3	2	1	2	3	4
25	0	1	1	1	1	1	1	1
26	0	0	0	0	2	2	2	2
27	0	0	1	2	1	1	1	1
28	2	1	1	0	1	1	1	1
29	0	0	1	1	0	0	0	0
30	0	0	0	0	0	0	0	0
31	0	1	0	1	1	1	1	1
32	0	0	0	1	1	0	1	1
33	0	0	0	0	0	0	0	0
34	0	0	0	0	0	0	0	0
35	0	0	1	0	0	0	0	0
36	0	0	0	0	0	1	1	1
37	0	0	0	0	1	1	1	0
38	0	0	0	0	0	0	0	0
39	0	0	0	0	0	0	0	0
40	0	0	0	0	0	0	0	1
X	3.1078	3.6115	3.9764	4.2257	4.4525	4.6283	4.8229	5.0019
σ	3.1312	3.6436	3.9930	4.1712	4.4345	4.6168	4.7750	4.9073

Table 4c. Distribution of Storm Duration (Hourly data: Greensboro, N.C., 1955-64) Storm Definition (x dry hours before storms)

	1	2	3	4	5	6	7	8
1	812	566	430	360	312	291	270	245
2	387	325	301	260	235	216	189	186
3	205	178	166	172	156	141	135	130
4	119	121	121	112	119	123	119	117
5	77	78	67	67	69	62	59	56
6	65	61	62	56	51	52	61	60
7	45	46	53	57	52	53	50	47
8	40	36	36	38	37	34	38	41
9	33	36	29	27	29	28	25	25
10	21	26	24	30	29	27	28	26
11	14	20	23	21	22	20	21	23
12	14	13	20	18	19	25	22	24
13	8	15	15	18	19	21	21	21
14	9	14	15	18	17	12	14	10
15	11	14	17	20	16	17	17	18
16	7	7	5	6	12	10	9	9
17	6	5	9	4	5	6	9	13
18	5	5	4	6	9	8	8	9
19	3	5	6	6	6	3	3	3
20	5	5	6	9	5	8	8	8
21	2	4	5	3	4	7	8	8
22	0	0	1	1	2	2	2	2
23	2	0	1	1	1	2	1	1
24	1	1	1	1	1	1	1	1
25	0	1	0	1	2	1	1	2
26	0	1	2	2	2	5	4	2
27	0	2	2	2	3	3	3	3
28	0	1	0	0	0	0	0	1
29	0	0	0	0	1	2	2	2
30	1	1	1	2	2	1	1	1
31	0	0	1	1	1	1	1	1
32	0	0	0	0	0	1	1	1
33	0	0	0	0	0	0	2	2
34	0	0	0	0	0	0	0	0
35	0	0	0	0	0	0	0	0
36	0	0	0	0	0	0	0	0
37	0	0	0	0	0	0	0	0
38	0	0	0	0	0	0	0	0
39	0	0	0	0	0	0	0	0
40	0	0	0	0	0	0	0	0
41	0	0	0	0	0	0	0	0
42	1	1	1	1	1	0	0	0
43	0	0	0	0	0	1	1	1
$\bar{X}$	3.1485	3.7500	4.1819	4.5114	4.8063	5.0296	5.2513	5.4186
σ	3.5462	4.1341	4.4792	4.7171	5.9789	5.2646	5.4408	5.5377

Table 5. Frequencies used and Parameter Estimates for L.N.M.

Place	Storm Def.	Cumulative Freq.	Estimated Parameters b	θ	a	χ^2	d.f.
Athens	1	1,4,10	1.22	0.917	1.63	13.5	11
	2	1,4,12	1.95	0.910	2.39	13.5	11
	3	1,3,16	2.29	0.917	2.76	9.1	12
	4	1,4,11	2.82	0.922	3.29	11.4	13
	5	1,2,11	3.44	0.924	3.90	7.7	13
	6	1,2,11	3.43	0.932	3.89	11.4	13
	7	1,2,15	3.64	0.933	4.10	11.5	14
	8	1,3,12	3.78	0.933	4.27	17.0	14
Charleston	1	1,3,7	1.47	0.849	1.98	15.7	10
	2	1,3,7	1.80	0.876	2.33	19.9*	11
	3	1,3,11	2.44	0.883	2.95	26.1**	13
	4	1,3,11	3.38	0.877	3.85	22.6*	13
	5	1,3,11	3.16	0.894	3.66	21.3	13
	6	1,3,5	3.11	0.896	3.67	21.7	13
	7	1,5,9	3.78	0.891	4.37	11.1	14
	8	1,3,5	3.89	0.895	4.52	14.4	14
Greensboro	1	1,3,11	1.56	0.898	1.63	10.5	13
	2	1,4,11	2.28	0.907	2.32	13.2	13
	3	1,3,14	2.59	0.909	2.75	21.6	14
	4	1,3,14	2.94	0.910	3.15	29.0**	15
	5	1,3,15	3.90	0.902	4.07	20.0	15
	6	1,3,14	4.06	0.915	4.17	21.3	15
	7	1,3,15	5.68	0.901	5.65	17.0	15
	8	1,3,15	5.47	0.908	5.54	26.8	17

(* Borderline at 5% level: ** Rejected at 5% Level.)

Table 6. Logarithmic Negative Mixture Distribution fitted to Selected Data

Place	Athens		Athens		Athens		Charleston		Charleston		Greensboro		Greensboro		Greensboro	
Storm Def.	1		4		5		1		7		1		6		8	
Variate (x)	Obs.	Comp.	Obs.	Comp.	Obs.	Comp.	Obs.	Comp.	Obs.	Comp.	Obs.	Comp.	Obs.	Comp.	Obs.	Comp.
1	363	363.0	163	163.0	139	139.0	538	538.0	208	208.0	812	812.0	291	291.0	245	245.0
2	291	275.7	170	168.5	148	148.0	419	455.7	227	239.6	387	383.3	216	210.6	186	182.3
3	142	142.4	114	108.1	102	99.6	274	237.3	182	161.3	205	208.7	141	146.4	130	133.7
4	71	85.9	67	74.4	71	70.9	157	140.5	109	113.9	119	128.9	123	106.3	117	101.0
5	51	56.7	50	53.8	54	52.6	87	89.8	80	83.2	77	86.1	62	79.9	56	78.1
6	38	39.8	32	40.3	32	40.1	49	60.5	58	62.5	65	60.5	52	61.5	60	61.4
7	32	29.1	29	31.0	29	31.4	40	42.2	49	47.9	45	44.2	53	48.3	47	49.0
8	25	22.0	25	24.4	24	25.0	27	30.3	43	37.3	40	33.2	34	38.6	41	39.6
9	16	17.0	24	19.5	24	20.2	21	22.2	27	29.4	33	25.5	28	31.2	25	32.3
10	16	13.4	19	15.8	19	16.5	16	16.5	20	23.5	21	19.9	27	25.5	26	26.6
11	12	10.8	19	13.0	15	13.7	13	12.5	14	18.9	14	15.8	20	21.1	23	22.1
12	14	8.7	13	10.8	15	11.4	7	9.5	13	15.3	14	12.6	25	17.5	24	18.4
13	10	7.2	9	9.0	9	9.6	8	7.3	16	12.5	8	10.2	21	14.7	21	15.5
14	8	5.9	6	7.6	8	8.1	{34	{27.6	11	10.31	9	8.3	12	12.4	10	13.0
15	{24	{34.2	8	6.4	7	6.9			12	8.5	11	6.9	17	10.5	18	11.0
16			9	5.5	10	5.9			10	7.1	7	5.7	10	8.9	9	9.4
17			{32	{35.6	{35	{39.5			7	5.9	{25	{30.4	6	7.6	13	8.0
18									{32	{32.0			8	6.6	9	6.9
19													{37	{42.5	8	5.1
20															{30	{38.1

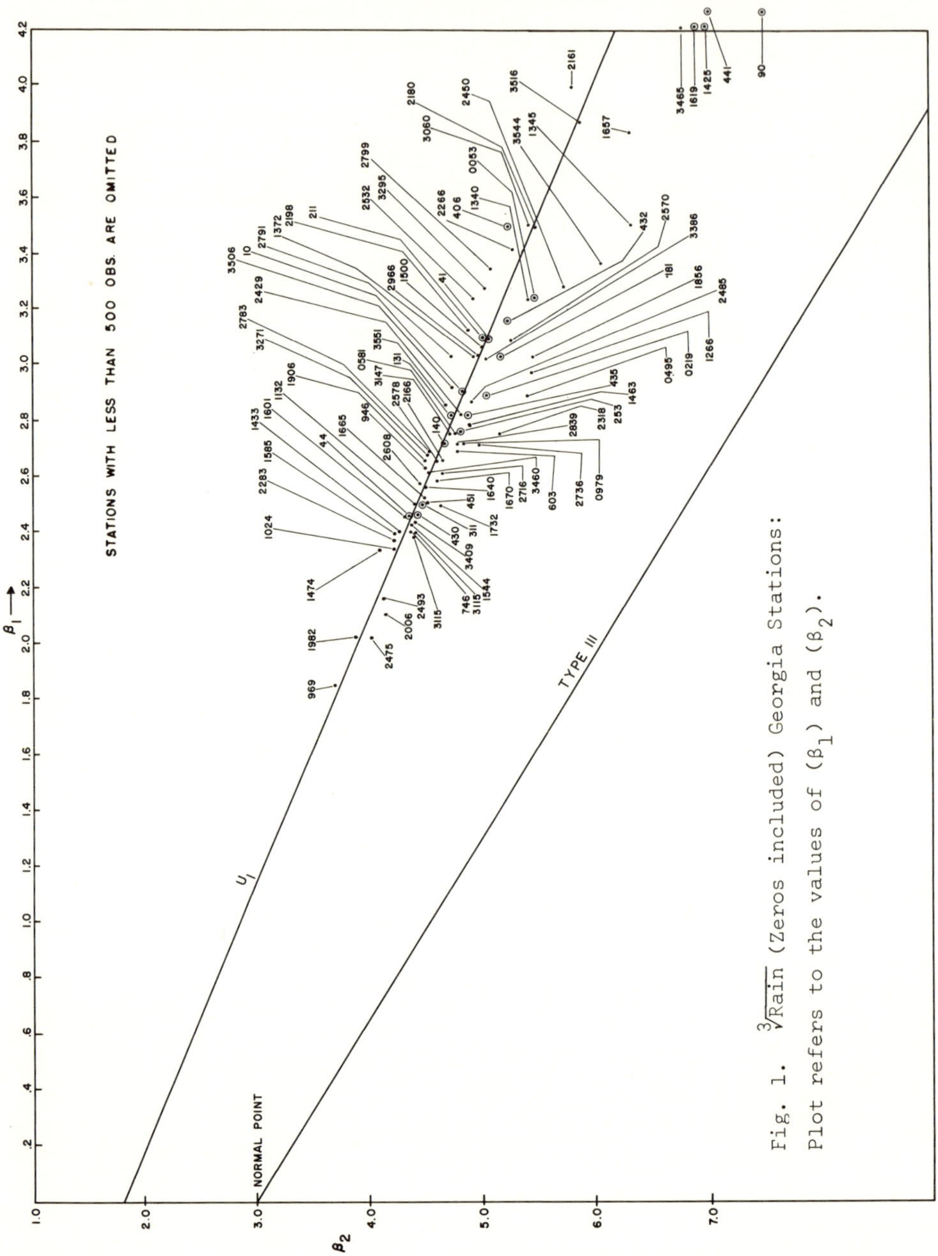

Fig. 1. $\sqrt[3]{\text{Rain}}$ (Zeros included) Georgia Stations: Plot refers to the values of (β_1) and (β_2).

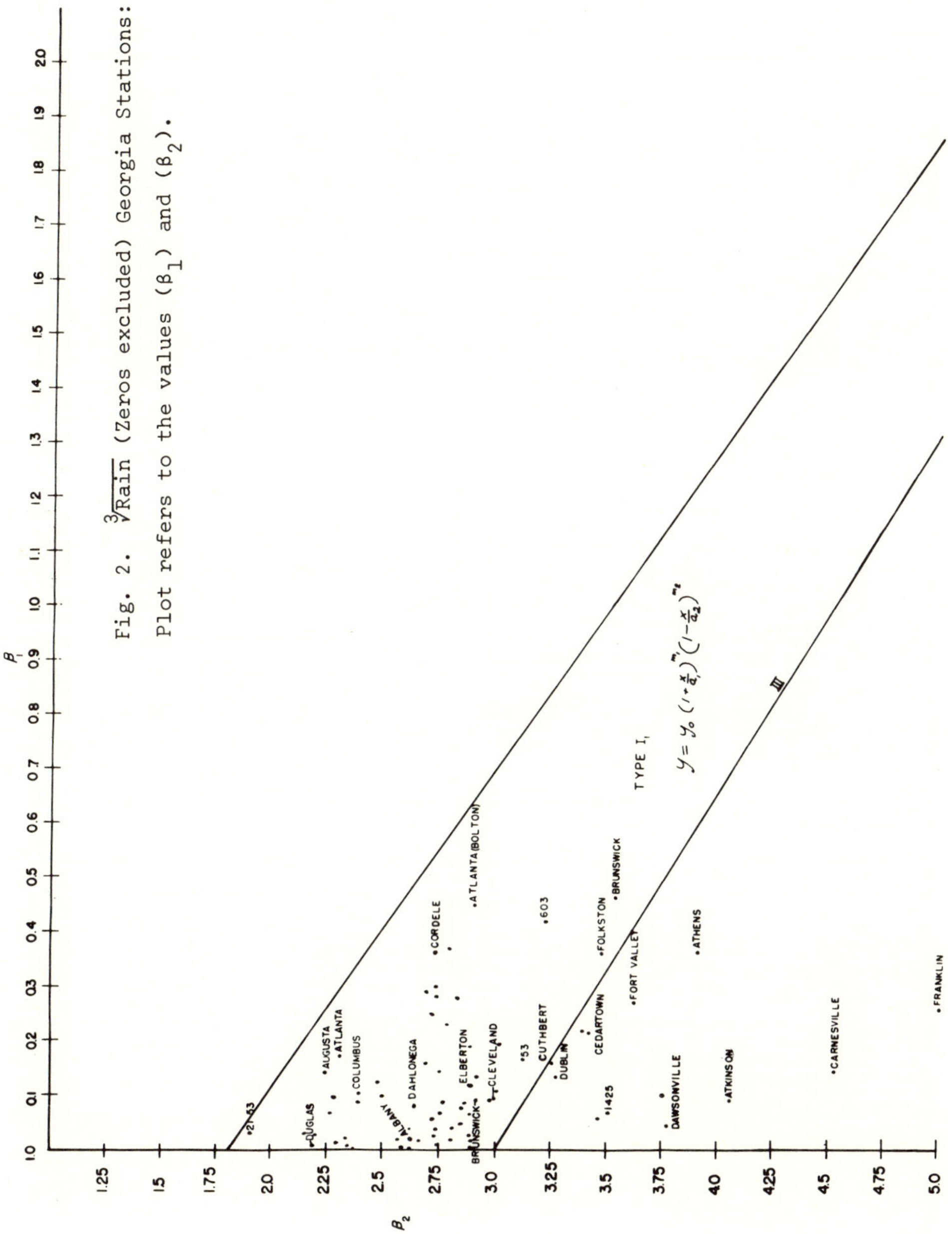

Fig. 2. $\sqrt[3]{\text{Rain}}$ (Zeros excluded) Georgia Stations: Plot refers to the values (β_1) and (β_2).

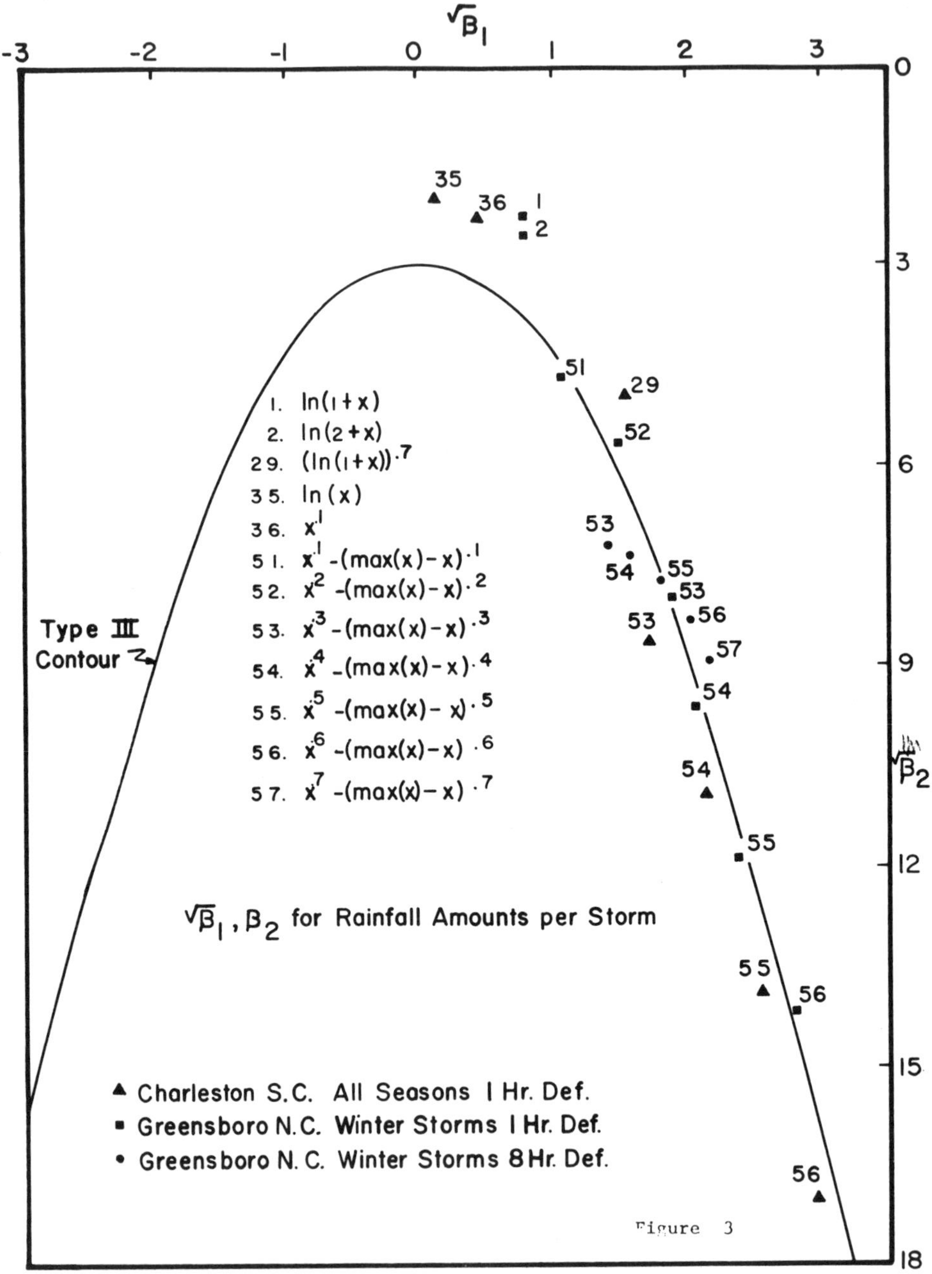

Fig. 3. $\sqrt{\beta_1}, \beta_2$ for Rainfall Amounts per Storm.

GROUP SIZE DISTRIBUTIONS, WITH APPLICATIONS TO VEHICLE OCCUPANCY

FRANK A. HAIGHT
The University of California at Los Angeles and
The Pennsylvania State University

SUMMARY

Group size distribution, occurring in the situation of vehicle occupancy, is studied by fitting Borel, Geometric and Poisson distributions to the actual data. Dual-occupancy model is discussed in detail and further generalizations are proposed.

1. INTRODUCTION

One curious aspect of group size distributions is the lack of a zero category. We can speak of the frequency of groups of size n for $n \geq 1$, but in every case $n = 0$ is without meaning. For example, it does not make sense to ask how often $n = 0$ animals are born in a litter, $n = 0$ cars are involved in a collision or $n = 0$ letters are mailed in a batch. This is in contrast to most well-known discrete distributions, such as the Poisson and binomial, which are intrinsically defined over $n = 0,1,2,\ldots$.

There are, of course, a number of simple methods for overcoming this difficulty. The best known is perhaps truncation, in which the p_0 value is simply prorated over the remining probabilities, giving a new distribution

$$q_n = \frac{p_n}{1 - p_0}, \quad n = 1,2,3,\ldots.$$

Another possibility is to reassign probabilities to cover the desired domain:

$$q_n = p_{n-1}, \quad n = 1,2,3,\ldots.$$

A variation of this method can be practiced on the data itself; for example, in considering vehicle occupancy, one may conveniently neglect the driver—a fixed feature of moving vehicles—and count, instead, only the passengers.

Finally, one may use distributions which naturally and by definition begin above zero. The Borel [1] distribution of the length of the discrete busy-period of a queue is a fairly well-known example, and one which will be used here:

$$q_n = \frac{n^{n-2}\, e^{-an}\, a^{n-1}}{(n-1)!}, \quad n = 1,2,3,\ldots.$$

Another distribution of this type is the Fisher logarithmic series:

$$q_n = \frac{-k^n}{n \log(1-k)}, \quad n = 1,2,3,\ldots.$$

In addition to these distributions, we will consider in the following section certain distributions which are too well known to require citation: Poisson, negative binomial, geometric, Neyman Types "A" and "C".

2. VEHICLE OCCUPANCY DATA

In considering measures for improving passenger transportation systems, one idea which appears particularly attractive, both by virtue of its simplicity and its economy, is that of increasing the utilization of established systems. In particular, it is often noted that private cars often carry far less than their capacity in many urban situations.

The extent of this underloading is still a matter of conjecture. Lauer [5] mentions taking some observations on the number of occupants per car on certain unspecified highways (presumably in Iowa), but does not give any results. It is often stated that cars in the United States carry on the average 1.5 occupants, without information either on the source of the data, or on the distribution. Mean values can sometimes be obtained from municipal traffic counts. For example, the City of Los Angeles cordon count records both vehicles and occupants at various locations, so that the mean values can be worked out.

During the spring and summer of 1959, the author, in an attempt to obtain more detailed information on this subject recorded vehicle occupancies of 14,876 passenger cars traveling the streets and freeways of Los Angeles. The information shown below is based on these samples, which were originally issued in an unpublished memorandum (Haight, [2]). Table 1 gives the basic information on the dates and locations of the samples, together with the mean and variance of total occupancy (including driver). In each case the sample given consists of a number of subsamples; for example, Sample 2 can be broken down into thirteen short counts taken on various freeways, and Sample 3 consists of subsamples of five-minute counts. The author will be glad to furnish interested readers with detailed information on these counts.

The means and variances of the samples and subsamples are shown in Figure 1, together with the appropriate theoretical curves for the Borel, Poisson, and geometric distributions. It will be noted that there is a considerable variability in the data, so that it would not be expected that all occupancy data could be fitted satisfactorily to a single distribution. On the whole, the geometric distribution seems best, especially with respect to the large samples.

In the report cited above, certain other results are given which are not relevant to counting distributions: occupancy by sex, diurnal trends, classification by economic area, comparison with traffic volume, etc. The results, however, in spite of the fairly large sample taken, are quite tentative on all of these points. The only tendency which seemed at the time to be well-established was a general increase in mean occupancy from 6:00 a.m. to 11:00 p.m., quite independently of the traffic volume peaks. (The approximate values were: 1.1 occupants/vehicle at 6:00 a.m., 1.7 occupants/vehicle at 1:00 p.m., 1.8 occupants/vehicle at 8:00 p.m., 2.2 occupants/vehicle at 11:00 p.m., 2.0 occupants/vehicle at 1:00 a.m.)

For fitting theoretical distributions, Sample 6 has been chosen as the most homogeneous. It consists of two parts: Sample 6A, passenger cars only, both east and west bound, corner of Wilshire and Bundy between 11:10 and 12:00 the morning of March 23, and Sample 6B, all vehicles at the same location the following day between 10:00 and 10:40.

In the results shown in Tables 2 and 3, the following principles were employed: occupants or passengers were counted as required by the domain of the distribution in question, and parameter estimation was by the method of moments. Values of Chi-square ranged from 6.89 for the two-parameter negative binomial for Sample 6A to the value of 199.88 for the Poisson. In every case the existing well-known distributions failed quite noticeably to account for the counts observed, and in many instances the most conspicuous failure was in the category of two occupants -- one passenger.

This simple observation, although quite understandable from the point of view of human behavior, does not contribute to any of the various deductive systems upon which the distributions under consideration have been based. It forms the basis of the generalization discussed in the following section.

3. A DUAL-OCCUPANCY MODEL

Suppose groups consist of two types of individuals (which for convenience we shall call men and women) indistinguishable in group size counting but obeying different laws of group formation. Specifically, assume that there is a fixed probability

β that a man in a group will bring one woman with him, and a probability $1 - \beta$ that he will not do so. Then if p_n represents the probability of n men in a group, and q_n the probability of n individuals in a group, it is not difficult to show that

$$q_1 = p_1(1 - \beta) ,$$
$$q_2 = p_2(1 - \beta) + p_1\beta ,$$
$$q_3 = p_3(1 - \beta)^3 + 2\, p_2\beta(1 - \beta) ,$$

and in general

$$q_n = \sum_{i=0}^{[\frac{1}{2}n]} \binom{n-1}{i} p_{n-i}(1 - \beta)^{n-2i} \beta^i, \quad n = 1,2,3,\ldots$$

where $[\frac{1}{2}n]$ denotes the greatest integer less than or equal to n. If we denote by m and v the mean and variance of the number of men in a group, and by M and V the corresponding quantities for total group size, we can obtain with some calculations the following relationships:

$$M = m(1 + \beta)$$
$$V = v(1 + \beta)^2 - m\beta^2 + m\beta \quad . \qquad (1)$$

If the generating function of q_n,

$$Q(s) = \Sigma\, s^n q_n ,$$

is formed, the coefficient of p_n will be found to be the binomial expansion of $s^n(s\beta - \beta+1)^n$; therefore

$$Q(s) = P[s(s\beta - \beta+1)].$$

If we assume some distribution for p_n which permits both m and v to be expressed in terms of a single parameter, then equations (1) will allow an evaluation by the method of moments of β and the parameter in the p_n distribution.

If, for example, influenced by Fig. 1, we choose the geometric distribution for groups of men:

$$p_n = (1 - r)\, r^{n-1} , \quad n = 1,2,\ldots,$$

the simultaneous solution of equations (1) yields, for sample 6A, the quadratic

$$3.10\, \beta^2 - 0.1205\, \beta - 0.1205 = 0$$

and the values

$$\beta = 0.2628, \quad r = 0.185 .$$

In this case the agreement is apparently superior to that obtained with other two-parameter distributions:

Occupants	Observed	Theoretical
1	902	882
2	403	435
3	106	102
4	38	35
5	16	11
$\geq$ 6	4	4

It would be interesting to compare this distribution systematically to large data samples in order to determine the precise situations in which it does give an improved fit. From a theoretical point of view, it would also be of interest to determine which distributions may be assumed for p_n. For example, Sample 6A, if taken with a Poisson p_n, yields a cubic equation in β which has no positive root. In fact, there will always be a cubic when p_n is chosen to be Poisson, and no doubt there exist other restrictions upon the choice of p_n.

4. GENERALIZATIONS

The model proposed* in Section 3 will yield a family of plausible group size counting distributions. Another such family of distributions is obtained from the "stuttering Poisson" model. In this system, groups of size n are assumed to arrive at instants which form a Poisson process. There are many applications (also with n continuous) of this scheme in insurance statistics (see, for example, Haight [3], Chapter 3.8).

The following generating-function relationship contains both the Hirata model of Section 3 and the Stuttering Poisson model:

$$Q(s) = P(sR(s)), \tag{2}$$

where P, Q and R are, respectively, the probability generating functions of three discrete distributions:

*Mr. Toshiro Edamura has informed me that this model appears in papers of Hirata [4] and Terao [6], both written in Japanese.

$$P(s) = \sum p_n s^n, \quad Q(s) = \sum q_n s^n, \quad R(s) = \sum r_n s^n .$$

Then the Stuttering Poisson probability q_n for n objects in unit time is obtained by setting

$$P(s) = \exp(-\lambda + \lambda s)$$

the Poisson generating function and assuming r_n to be any group size distribution. For particular choices of r_n, one obtains a Laguerre Polynomial (the 'Galliher process'), the negative binomial distribution (the 'Arfwedson process') and the Neyman Type A distribution. Details are given by Haight [37].

The Hirata model is obtained from Eq. (2) by setting

$$R(s) = 1 - \beta + \beta s$$

The argument leading to Eq. (2) can be summarized as follows: let p_n represent the probability of n <u>primary</u> objects in a group, and r_n the probability that a primary object is accompanied by n <u>secondary</u> objects. Then the generating function for the cluster of a single primary object and its associated secondary objects is sR(s). Therefore the distribution of total group size is convolution-mixed with respect to p_n:

$$q_n = \sum p_n^{j*} h_j \tag{3}$$

where the probability generating function of the h_j is sR(s). Eq. (2) is now easy to obtain from Eq. (3).

There are also, quite naturally, many other interesting special cases of Eq. (2). For example, if p_n is geometric in the Hirata model, q_n resembles a Chebyshev polynomial, the only exception being that the successive terms do not alternate in sign.

The relationships between the means $\underline{a}$, $\underline{b}$, and $\underline{c}$ and the variances $\underline{A}$, $\underline{B}$, and $\underline{C}$ of p_n, q_n, and r_n, respectively, are easy to obtain from Eq. (2) and are as follows:

$$b = a(1 + c)$$

$$B = aC + A(1+c)^2,$$

from which the corresponding formulas (Eq. (1)) of Section 3 follow.

Generalizations of this scheme could very well be based on successive 'generations' of secondary, tertiary, etc., objects, or on secondary objects of a variety of types.

REFERENCES

[1] Borel, E. 1942. Sur l'emploi du theorème du Bernoulli pour faciliter le calcul d'un infinité de coefficients. Application au problème de l'attente a un guichet. C. R. Acad. Sci., Paris 214:452-6.

[2] Haight, Frank A. 1959. Statistical Studies in California Traffic I. Vehicle Occupancy in Los Angeles (unpublished).

[3] ————. 1967. Handbook of the Poisson Distribution, John Wiley, New York.

[4] Hirata, Shinzo. 1933. The movement of people at the ticket office of Shinzyuku Station. Kagaku (Science) 3 (No. 3) (In Japanese).

[5] Laver, A. R. 1953. A sampling study of drivers on the highways for the 24-hour period. Highway Research Board Bulletin No. 73. National Academy of Sciences, Washington, D. C.

[6] Terao, Senzo. 1949. On the distribution of combined pedestrians. Oyobuturi (J. Applied Physics) 18 (No. 4) (In Japanese).

Table 1. Summary of Samples and Results

Sample No.	No. of Vehicles	Day and Date	Time	Location	Mean Occupancy	Variance of Occupancy
1	1200	Thurs. Mar. 19	Morn. & Aft.	Sepulveda	1.4458	.6271
2	1371	Mon.,Tues.July 13,14	Afternoon	Freeways	1.5270	.9460
3	1193	Wed. July 15	Afternoon	La Cienega-Beverly	1.4510	.8040
4	2189	Thurs. July 16	Afternoon	La Cienega-Beverly	1.48	.746
5	245	Thurs. July 16	Afternoon	La Cienega-Beverly	(Classified by Sex)	
6	2480	Mon.,Tues.Mar.23,24	Morning	Wilshire-Bundy	1.52	.722
7	400	Wed. Mar. 25	Morning	Various	1.43	.7501
8	3095	Fri-Sun.Mar.13,14,15	Various	L. A. Int. Airport	1.807	1.416
9	2703	Fri.,Sat.Aug. 7-8	All Day	Wilshire-Wellesley	1.7440	.9911
Total	14876	1959		Los Angeles	1.606*	.97642*

*Excluding Sample 5.

Table 2. Various Distributions Fitted to Sample 6.A

Occupants	Passengers	Observed	Geometric	Borel	Negative Binomial	Fisher Series	Neyman Type A	Neyman Type C	Poisson
1	0	902	945	1204	909	787	915	913	845
2	1	403	337	257	381	298	367	370	467
3	2	106	120	97	126	150	133	130	129
4	3	38	43	43	38	68	40	38	24
5	4	16	15	21	11	51	14	18	3
6 or more	5 or more	4	8	27	4	114			—

Table 3. Various Distributions Fitted to Sample 6.B

Occupants	Passengers	Observed	Negative Binomial	Fisher Series	Neyman Type A	Borel	Geometric
1	0	678	685	677	699	720	665
2	1	227	206	195	175	174	227
3	2	56	74	75	86	63	78
4	3	28	28	33	33	27	27
5	4	8	11	15	18	13	9
6 or more	5 or more	14	7	15		14	5

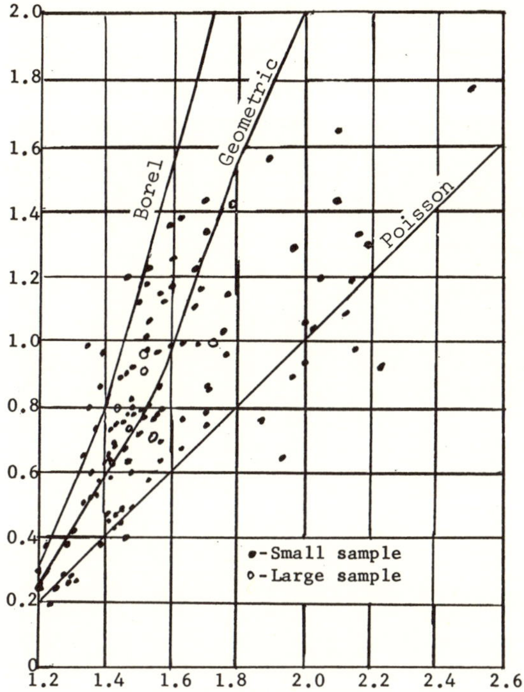

Mean Occupancy

Fig. 1. Variation in traffic volume and mean occupancy during a 20-hour period.

THE CLASSICAL OCCUPANCY PROBLEM REVISITED

WILLIAM L. HARKNESS
The Pennsylvania State University
Department of Statistics

SUMMARY

If r balls are distributed randomly in n cells, then expressions for the probability $P_x(r,n,1)$ that exactly x cells remain empty are well-known. More generally, if r shots are fired randomly at n targets, with the probability being p that any specified shot hits its target, then the probability $P_x(r,n,p)$ that exactly x targets are missed (so that n-x are hit) is given by

$$P_x(r,n,p) = \binom{n}{x} \sum_{k=o}^{n-x} (-1)^k \binom{n-x}{k} \left[1- \frac{(\nu+x)p}{n}\right]$$

Derivations and properties of these "classical occupancy" distributions are fairly well-known; some applications have been given. Further applications of these distributions are presented, for example, to the problem of determining the number x of survivors when n offensive units are attacked by r defenders, each having probability p of successfully destroying a (randomly chosen) offensive unit. A probability model in which the set of probabilities $\{P_x(r,n,p)\}$ serve as (stationary) transition probabilities in a Markov chain is discussed. Additional examples of random phenomena having probability distributions given by $P_x(r,n,p)$ are considered.

1. INTRODUCTION

If r balls are randomly distributed into n cells (so that each ball has probability 1/n of falling into the ith cell, $i = 1,\ldots,n$) then it is well known that the probability that there are exactly x empty cells (assuming no restrictions on the number of balls per cell) is

$$P_x(r,n,1) = \binom{n}{x} \sum_{k=o}^{n-x} (-1)^k \binom{n-x}{k} [1- \frac{k+x}{n}]^r, \quad x = 0,\ldots,n-1 \qquad (1)$$

(of course $P_n(r,n,1) = 0$). Now suppose this model is modified in the following way. As before, we assume that each ball has probability 1/n of falling into the ith cell; however, the ball now is presumed to have probability p of staying in the cell and probability (1-p) of falling through (and hence is not available to "fill" a cell). In this case, the probability distribution of the number of empty cells may be shown (e.g., by using the method of "inclusion and exclusion"; cf. Feller [11], p. 106) to be given by

$$P_x(r,n,p) - \binom{n}{x} \sum_{k=0}^{n-x} (-1)^k \binom{n-x}{k} [1- \frac{(k+x)p}{n}]^r, \quad x = 0,\ldots,n \qquad (2)$$

(note that in this case $P_n(r,n,p) = (1-p)^n$). These probabilities satisfy the recursion

$$P_x(r+1,n,p) = (1-px/n)P_x(r,n,p)+p(x+1)/nP_{x+1}(r,n.p). \qquad (3)$$

The validity of (3) may be reasoned as follows: the left-hand side is the probability of the event that there are exactly x empty cells after r+1 balls are distributed over the n cells. This event will be satisfied if, after r balls have been distributed among the n cells, there are either (a) x unoccupied cells (occurring with probability $P_x(r,n,p)$) and the $(r+1)^{st}$ ball <u>does not</u> come in contact and remain in one of these empty cells (an event having probability 1-px/n); or (b) (x+1) unoccupied cells (occurring with probability $P_{x+1}(r,n,p)$) and the $(r+1)^{st}$ ball <u>encounters</u> and <u>remains</u> in one of these (x+1) cells (an event having probability p(x+1)/n), to leave x empty cells. The events in (a) and (b) are mutually exclusive, and hence the probability that exactly one occurs is given by the right-hand side of (3).

In view of the nature of phenomenon of this type (and in keeping with previous historical descriptions), it is natural to refer to the three-parameter family of distributions given by (2) as "classical occupancy distributions with parameters r, n, and p" (or CLOD's, for short!). Our interest here resides primarily in enumerating various applications and physical models for which CLOD's with $p<1$ are appropriate. Numerous examples of situations for which CLOD's with $p = 1$ occur have been amply described and thus need not be articulated here. It suffices to note that Feller [11] mentions no less than sixteen such applications ranging from such diverse fields as biology to nuclear physics.

2. APPLICATIONS AND PHYSICAL MODELS

(a) One of the earliest (if not the first) derivations of (2) was given by Sprott [19] in deriving the probability distribution of the number of targets hit when r shots (balls) are fired randomly at n targets (cells) and there is a probability p that any specified shot hits the target. The derivation parallels exactly that given by Feller ([11], p. 101-2) for the case $p = 1$ except that (in Feller's notation)

$$p_i = (1-p/n)^r, \; p_{ij} = (1-2p/n)^r, \; p_{ijk} = (1-3p/n)^r, \ldots,$$

where $p_i = P(A_i)$, $p_{ij} = P(A_iA_j)$, $p_{ijk} = P(A_iA_jA_k)$, and A_α is the event that cell number α is empty ($\alpha = 1,\ldots,n$). One clearly arrives at (2) in the same manner that Feller arrives at (2.4) (see p. 102). Sprott also noted that modifications of this formula have been used in certain biological problems (e.g., Rashevsky [18] considers the problem of assigning biological functions to cells).

(b) Barton and David [3] describe a model for (2) in considering what they termed "the Golliwog Problem." Consider children (or any appropriate group!) who buy and eat a certain brand of marmalade (sic). With each jar of marmalade there is given away a paper "golliwog," which may be small or large. The large species are the same. The small species is composed of n distinct varieties. The child's task is to collect d of the large species and one each of the n different small kinds. If r jars of marmalade are purchased, the probability that there are exactly x "small" species of golliwogs missing is given by

(2) (where, say, n and p are determined by the company processing the marmalade--one could, e.g., set p = n/n+c, where c is arbitrary, perhaps equal to one). This model embodies many of the ingredients typical of coupon-collections.

(c) In sampling theory, consider a population divided by some criterion into n groups. Suppose a group is selected at random and a member of the group is drawn randomly. Further, assume that a certain characteristic of the population occurs with probability p. For example, the "groups" could be the provinces, states, or some other well-defined region of a country and the characteristic could be a "skin-color," "albinism," "heart trouble," etc., occurring with (assumed) constant probability p in each region. If a random sample of r individuals is drawn, one from each group, then the probability that in exactly x of the groups no member with the given characteristic is found is given by (2).

(d) In "air-battle" theory (or predator-prey models) one identifies the n cells with "bombers" or "missiles" (or more generally, with "offensive forces") and the r balls with "fighters" or "anti-missiles" (or more generally, with "defensive forces"). The fighters "intercept" or "defend against" the bombers, engaging in an air battle. In this instance $P_x(r,n,p)$ represents the probability that precisely x of the n bombers survive the air battle when intercepted by r fighters; p is a "kill" parameter corresponding to a probability that a fighter is able to intercept and destroy a bomber. More generally, an air battle may be considered as a sequence of "sub-battles" in each of which a wave of interceptors attacks the bombers that survived the previous sub-battle. Thus, in order that a flight of n bombers reach their destination, it may be necessary that they successively pass over k defensive bases. It is usually assumed that during a given sub-battle, when one interceptor wave is defending, all of the "dying" bombers crash. With this type of model, it is natural to use a Markov chain with (perhaps stationary) transition probabilities $P_{ij} = P_j(r,i,p)$ to represent the total history of the process. Occasionally, it is desirable to permit the number of defensive units to vary from one sub-battle to the next, so that the transition probabilities in the mth sub-battle, namely $P_{ij}(m)$, may depend on m through $r = r_m$.

In biological models, this "fighter bomber" model has a

"predator-prey" interpretation. Thus, Mertz and Davies [13] develop a model describing the number of pupal prey surviving the "vulnerable period." All adults are assumed to be equally predatious and to act independently of one another in their predatory behavior. The pupal prey and prey corpses are assumed to be equally and synchronously vulnerable. A prey which has been attacked once is regarded as a corpse; i.e., no provision is made for mortality associated with cannibalism or from the survival of wounded pupae. The corpses are assumed to be capable of sustaining an unlimited number of attacks during the collective period of vulnerability. The authors find that, experimentally and theoretically, the number of prey surviving (at time t, say) has a distribution given by (2).

Finally, in the same context, Peto [16] analyzes (in detail) a dose-response equation for the invasion of micro-organisms. A series of k doses ($r_1,\ldots,r_k$ micro-organisms) is administered to varying numbers of $m_1,\ldots,m_k$) test animals, and the corresponding survivors are observed to number $x_1,\ldots,x_k$. Peto assumes that (i) the test animals are homogeneous, (ii) the probability of one organism killing its host is p (small) and (iii) the organisms act independently of each other. Thus, for a dosage of r_1 micro-organisms to n = 1 animal, we have an (in a sense, trivial) application of (2) with parameters $r = r_1$, n = 1, and p. Nevertheless, the model is extremely important. Peto's main concern is the estimation of p.

(e) In a theoretical investigation of the spread of epidemics by carriers, Weiss [23] considers a "closed population of n susceptible individuals (to a fixed disease) and lets r be the number of carriers (of the same disease). The diseases may, for example, be typhoid, amoebic dysentery, or typhus, while the "carrier" could be a human or a polluted stream. A carrier is defined to be an individual who does not have overt disease symptoms but who can transmit the disease. At a fixed time t, assume that the probability that a "carrier" is eliminated in $(t,t+h)$ is βrh, where $r = r(t)$ is the number of carriers extant at time t. The probability of a carrier infecting a susceptible in $(t,t+h)$ is αrh; α and β are constants. It is assumed that a susceptible that has been infected will <u>not</u> be a further source of infection, but will be identified and removed from the population. This model leads to a random walk on a rectangular lattice with points (r,n) with transitions and corresponding probabilities given by

$$P\{(r,n) \rightarrow (r,n-1)\} = n\sigma/1+n\sigma; \tag{4}$$

$$P\{(r,n) \rightarrow (r-1,n)\} = 1/1+n\sigma,$$

where $\sigma = \alpha/\beta$ represents the ratio of the rate of infection to the rate of elimination. In this situation, let $P_x(r,n,\sigma)$ be the probability that a population of n susceptibles is reduced to x survivors at the termination of an epidemic initiated (at time t) by r carriers. Weiss then argues (analogously to (3)) that

$$P_x(r,n,\sigma) = \{P_x(r-1,n,\sigma)+n\sigma P_x(r,n-1,\sigma)\}/(1+n\sigma); \tag{5}$$

this recurrence relation is then used to solve for $P_x(r,n,\sigma)$ explicitly. Weiss finds the solution

$$P_x(r,n,\sigma) = \binom{n}{x}\sum_{k=o}^{n-x}(-1)^k\binom{n-x}{k}\{1+(k+x)\sigma\}^{-r}, \quad x = 0,1,\ldots,n. \tag{6}$$

In order that (6) represent a probability distribution, it is sufficient that σ and r be non-negative real numbers. Further, it is seen that (6) can be viewed as an extension of (2), through replacing (r) by (-r) and p/n by $-\sigma$. In fact, it follows from (2) and (6) that the family of functions of the form given by (2) is, for each positive integer n, a probability distribution (i) for all real r>o and $o \leq p \leq 1$ and (ii) for all negative real numbers r and p. Both of the recurrence relations (3) and (5) are satisfied by all members of this family of distributions.

3. STATISTICAL CHARACTERISTICS

Let x be a random variable (r.v) with probability density function (p.d.f.) $P_x(r,n,p)$ of the form given in (2) being nonzero only for x = 0,1,...,n, and defined for (i) r>o and $o \leq p \leq 1$ and (ii) r<o and p<o. In this section we derive some of the more important statistical properties of this family.

(i) The probability generating function g(r,n,p,t) of x is given by

$$g(r,n,p,t) = \sum_{x=o}^{n} P_x(r,n,p)t^x = \sum_{x=o}^{n}\binom{n}{x}\sum_{k=o}^{n-x}(-1)^k\binom{n-x}{k}[1-\frac{(k+x)p}{n}]^r t^x$$

$$= \sum_{x=o}^{n}\binom{n}{x}\sum_{\nu=x}^{n}(-1)^{\nu-x}\binom{n-x}{\nu-x}[1-\frac{\nu p}{n}]^r t^x$$

$$= \sum_{\nu=0}^{n} \binom{n}{\nu} (1- \frac{\nu p}{n})^r \sum_{x=0}^{\nu} (-1)^{\nu-x} \binom{\nu}{x} t^x$$

$$= \sum_{\nu=0}^{n} \binom{n}{\nu} (1- \frac{\nu p}{n})^r (t-1)^\nu = \sum_{\nu=0}^{n} (n)\nu \, (1- \frac{\nu p}{n})^r \frac{(t-1)^\nu}{\nu!} \qquad (7)$$

From the form of g(r,n,p,t) as given by (7), we see that the νth factorial moment $E[x(x-1)\ldots(x-\nu+1)] = E[(x)\nu] = \mu_{[\nu]}$ is given by

$$\mu'_{[\nu]} = (n)_\nu \, (1- \nu p/n)^r \qquad (8)$$

(ii) The moment generating function $M_x(r,n,p,t)$ of x is given by

$$M_x(r,n,p,t) = g(r,n,p,e^t) = E(e^{tx})$$

$$= \sum_{\nu=0}^{n} (-1)^\nu \binom{n}{\nu} (1- \frac{\nu p}{n})^r \sum_{j=0}^{\nu} (-1)^j \binom{\nu}{j} e^{tj}$$

$$= \sum_{m=0}^{\infty} \left\{ \sum_{\nu=0}^{n} \binom{n}{\nu} (1- \frac{\nu p}{n})^r (-1)^\nu \sum_{j=0}^{\nu} (-1)^j \binom{\nu}{j} j^m \right\} \frac{t^m}{m!}$$

$$= \sum_{m=0}^{\infty} \mu'_m \, t^m/m!, \qquad (9)$$

where

$$\mu'_m = \sum_{\nu=0}^{m} \binom{n}{\nu} (1- \frac{\nu p}{n})^r \Delta^\nu 0^m = E(x^m) \qquad (10)$$

is the mth moment about the origin of x. The quantity $\Delta^\nu 0^m = (-1)^\nu \sum (-1)^j \binom{\nu}{j} j^m$ is the νth order difference of zero of the mth order and vanishes for $\nu > m$ (Patil and Joshi [15], pp. 2 and 56). In particular, from either (8) or (10) we can find that the mean μ'_1 and variance σ^2 of x are given by

$$\mu'_1 = E(x) = n(1-p/n)^r$$

and

$$\sigma^2 = n(n-1)(1-2p/n)^r + n(1-p/n)^r - n^2(1-p/n)^{2r} \qquad (11)$$

The moments μ'_m bear a striking resemblance to the corresponding moments of the binomial distribution with parameters n and $\theta = (1-p/n)^r$. The νth factorial moment of a binomial distribution with these parameters is known to equal to $(n)_\nu \theta^\nu$, which differs but slightly from the expression given in (8). Proceeding in a manner paralleling the derivation of (9) and (10), the mth moment (say a'_m) of the binomial distribution is found to be equal to

$$a'_m = \sum_{\nu=o}^{m} \binom{n}{\nu} \theta^{\nu} \Delta^{\nu} o^{m} \tag{12}$$

for all positive integers ν and m, $\Delta^{\nu}o^{m} \geq o$. From the inequalities

$$\text{(i)} \quad (1-\nu p/n)^{r} \leq (1-p/n)^{\nu r} = \theta^{\nu}, \text{ if } r>o, \ o \leq p \leq 1$$

$$\text{(ii)} \quad (1-\nu p/n)^{-r} \geq (1-p/n)^{-\nu r} = \theta^{\nu}, \text{ if } r<o, \ p<o$$

and the fact that $\Delta^{\nu}o^{m} \geq o$ for all positive integers ν and m, we deduce that

$$\mu'_m \leq a'_m, \ m = 1,2,\ldots, \text{ if } r>o \text{ and } o \leq p \leq 1,$$

while (13)

$$\mu'_m \geq a'_m, \ m = 1,2,\ldots, \text{ if } r<o \text{ and } p<o.$$

It is also worth noting that the probability density function $P_x(r,n,p)$ also strongly resembles the binomial distribution with these parameters; indeed

$$b(x;n,\theta) = \binom{n}{x} \theta^{x}(1-\theta)^{n-x} = \binom{n}{x} \sum_{k=o}^{n-x} (-1)^{k}\binom{n-x}{k}[1-p/n]^{(k+x)r}$$

$$\simeq \binom{n}{x} \sum_{k=o}^{n-x} (-1)^{k}\binom{n-x}{k}[1-(k+x)p/n]^{r} = P_x(r,n,p). \tag{14}$$

Clearly, a binomial approximation to $P_x(r,n,p)$ is suggested by the above observations; the adequacy of its use may be determined by reference to Tables I, III, IV and VI where values of a Poisson approximation are also given. For these values of r,n, and p, it is seen that the binomial approximation (given in (14)) is vastly superior to that given by the Poisson. As in Feller [11], pp. 103-5, it can be shown that $P_x(r,n,p)$ converges to the Poisson distribution with parameter λ as r and $n \to \infty$ in such a way that $n(1-p/n)^{r} \to \lambda$, $o<\lambda<\infty$. Finally, using methods similar to those of Weiss [24], for instance, it can be shown that the standardized random variable $z = (x-\mu'_1)/\sigma$ has a limiting normal distribution with mean zero and variance one as r, n, and $n(1-p/n)^{r} \to \infty$.

4. MARKOV CHAINS

It has been observed previously that certain physical models lead to a study of Markov chains with (stationary or otherwise) transition probabilities $P_{ij} = P_j(r,i,p)$. Since $P_{ij} = o$ for $j>i$ and $P_{oo} = 1$, in this case, these chains are "irreversible" with one absorbing state--only transitions "towards" the absorbing 0 are possible (i.e., a transition to a state i after a passage from this state to another state $j<i$ is impossible). Let X_k denote the (random) state of the chain after k transitions. Thus, in the "airbattle" model, X_k denotes the number of bombers (initially, $x_o = n$ in number) surviving the kth sub-battle. In the succeeding discussion, we focus attention on such quantities of interest as the expected state $E(X_k)$ of the process and the probability of the process being in state x after k transitions. Throughout the following discussion, let the initial state be $x_o = n$, with probability one.

Letting $A = (P_{ij})$ be the matrix of transition probabilities (so that A is lower triangular), the k-step transition probabilities $P_{ij}^{(k)}$ are the elements of the kth power of A. Of course, these transition probabilities satisfy the relations

$$P_{ij}^{(k)} = \sum_{\alpha=j}^{\alpha=i} P_{i\alpha}^{(k-1)} P_{\alpha j}, \quad k = 2,3,\ldots.$$

Explicit "closed" form expressions for these sums of a simpler nature do not seem to be possible to obtain. We consider two approximations for these k-step transition probabilities.

The first approximation is obtained by simply replacing the exact values $P_j(r,i,p)$ in A by the binomial approximating values

$$\tilde{P}_j(r,i,p) = \begin{cases} \binom{i}{j} \theta_i^j (1-\theta_i)^{i-j}, & j = o,1,\ldots,i;\ i = 1,2,\ldots \\ o & j = i+1,i+2,\ldots \end{cases} \tag{15}$$

where $\theta_i = (1-p/i)^r$, $i = 1,2,\ldots$ (with, of course $\tilde{P}_o(r,o,p) = 1$, $\tilde{P}_j(r,o,p) = o$ for $j>o$. Note that for $k = 2$, we have

$$\tilde{P}_{ij}^{(2)} = \sum_{\alpha=j}^{i} \tilde{P}_{i\alpha} \tilde{P}_{\alpha j} = \sum_{\alpha=j}^{i} \binom{i}{\alpha} \theta_i^{\alpha} (1-\theta_i)^{i-\alpha} \binom{\alpha}{j} \theta_\alpha^j (1-\theta_\alpha)^{\alpha-j}, \tag{16}$$

which does not reduce to a tractable simpler form either.

This approximation, then, would appear to be of little value, in as much as exact evaluation (by high-speed computers) presents little more difficulty than that required using the given approximation.

The second approximation is motivated by the following considerations. The expected value $E(X_2)$ of X_2 (understanding that $x_o = n$ with probability one) can be written more suggestively in the form

$$E(X_2) = \sum_{\alpha=o}^{n} \alpha P_{n\alpha}^{(2)} = \sum_{\alpha=1}^{n} E(X_2|X_1 = \alpha)P\{X_1 = \alpha\}$$

$$= \sum_{\alpha=1}^{n} \alpha(1-p/\alpha)^r P_{n\alpha} = E[\varphi(X)] \qquad (17)$$

where $\varphi(X) = X(1-p/X)^r$, and the last expectation is taken with respect to the distribution $\{P_{n\alpha}\}$ of X. It is readily verified that φ is a convex function of x for $x \geq p$ (since $\varphi''(x) \geq o$ for $x > p$), so that by Jensen's inequality

$$E(X_2) = E[\varphi(X)] \geq \varphi[E(X)] = n\delta_2^{(n)}, \qquad (18)$$

where $\delta_2^{(n)} = \lambda_1^{(n)} \lambda_2^{(n)}$, $\lambda_1^{(n)} = (1-p/n)^r$, and $\lambda_2^{(n)} = (1-p/n\lambda_1^{(n)})^r$. The validity of (18) requires that $n\lambda_1^{(n)} > p$. We then approximate $P_{ij}^{(2)}$ by $b_{ij}^{(2)}$ where

$$b_{ij}^{(2)} = \binom{i}{j}[\delta_2^{(i)}]^j[1-\delta_2^{(i)}]^{i-j}, \quad j = o,1,\ldots,i. \qquad (19)$$

Numerical evidence indicates that this binomial approximation is surprisingly adequate even for very small values of r and n. If $i\lambda_1^{(i)} \leq p$, the approximation in (19) breaks down; a practical solution is to assume that in this case $\delta_2^{(i)} = o$, so that we put $b_{io}^{(2)} = 1$ and $b_{ij}^{(2)} = o$ for $j > o$. Proceeding inductively, it follows from repeated applications of the arguments above that

$$E(X_k) \geq n\delta_k^{(n)}, \text{ for } k = 1,2,\ldots, \qquad (20)$$

where

$$\delta_k^{(n)} = \lambda_1^{(n)} \lambda_2^{(n)} \ldots \lambda_k^{(n)}, \text{ and } \lambda_m^{(n)} = (1-p/n\,\delta_{m-1}^{(n)})^r,$$

for $m = 2,3,\ldots$, provided that $n\delta_{k-1}^{(n)} \geq p$. One then approximates $P_{ij}^{(k)}$ by binomial probabilities $b_{ij}^{(k)}$ depending on the parameters i and $\delta_k^{(i)}$.

Upper bounds for $E(X_k)$ can also be derived simply. The inequality $(1-p/\alpha)^r$ holds for all permissible values of r,n,p (i.e., for which $P_x(r,n,p)$ is nonzero) if $\alpha \leq n$. Noting (17), it follows that

$$E(X_2) \leq (1-p/n)^r \sum_{\alpha=1}^{n} \alpha P_{n\alpha} = n(1-p/n)^{2r} . \tag{21}$$

Repeating this argument, we obtain the inequality

$$E(X_k) \leq n(1-p/n)^{kr}, \; k = 1,2,\ldots. \tag{22}$$

The inequalities (20) and (22) provide upper and lower bounds for $E(X_k)$, which for moderate values of r and n seem to delimit $E(X_k)$ quite satisfactorily.

5. MIXTURES

If X is a random variable with discrete probability density function $P(x|\theta)$, where the parameter θ is a possible value of a random variable Ⓗ having p.d.f. $f(\theta)$, then the marginal distribution $P(x)$ of X is said to be a mixture on θ with mixing distribution $f(\theta)$. In the present situation we consider mixtures on r of CLOD's in which the mixing distributions are (i) Poisson, (ii) binomial, and (iii) negative binomial. Let R denote the r.v. having values $r = 0,1,2,\ldots$.

(i) (Poisson mixtures). Assume that the distribution of R is Poisson with parameter λ, i.e.,

$$f^{(1)}(r,\lambda) = P\{R = r\} = e^{-\lambda} \lambda^r/r!, \; r = 0,1,\ldots.$$

Then the corresponding mixture $P_x^{(1)}(\lambda,n,p)$ is given by

$$P_x^{(1)}(\lambda,n,p) = \sum_{r=o}^{\infty} P_x(r,n,p) \; f^{(1)}(r,\lambda)$$

$$= \binom{n}{x} \sum_{j=o}^{\infty} (-1)^j \binom{n-x}{j} e^{-\lambda} \sum_{r=o}^{\infty} \{\lambda[1-(j+x)p/n]\}^r$$

$$= \binom{n}{x} \sum_{j=o}^{n-x} (-1)^j \binom{n-x}{j} \; \theta^j \theta^x = \binom{n}{x} \; \theta^x (1-\theta)^{n-x}, \tag{23}$$

where $\theta = e^{-\lambda p/n}$. Thus, for example, if r balls are randomly distributed in n cells with the number of balls to be distributed having a Poisson distribution, then the number of empty cells has a binomial distribution.

(ii) (Binomial mixture). The distribution of R is assumed to be given by

$$f^{(2)}(r,\lambda) = \binom{N}{r}\lambda^2(1-\lambda)^{N-r}, \quad r = 0,1,\ldots,N.$$

Easy calculations show that the mixture

$$P_x^{(2)}(\lambda,n,p) = P_x(N,n,\lambda p) \tag{24}$$

(iii) (Negative binomial mixture). Assuming that R has the negative binomial distribution

$$f^{(3)}(r,\lambda) = \binom{\alpha+r-1}{\alpha-1}(\lambda/1+\lambda)^r(1/1+\lambda)^\alpha, \quad r = 0,1,2,\ldots$$

we obtain

$$P_x^{(3)}(\lambda,n,p) = P_x(\alpha,n,-\lambda p) \tag{25}$$

It is of some interest to note these interrelations (23), (24), (25) between the Poisson, binomial, negative binomial, and CLOD's, together with the results on the various approximations.

6. DESCRIPTION OF TABLES

Tables I, III, IV, and VI give values of the exact distribution $P_x(r,n,p)$. (1st line in each group of three lines), the binomial approximation (15) (2nd line in each group of three), and the Poisson approximation (with parameter $\lambda = ne^{-rp/n}$) in the 3rd line of each group. Thus, in Table I, for n = 2, we have

$$P_o(10,2,.10) = 0.15120, \quad P_1(10,2,.10) = 0.50012, \text{ etc. (exact)}$$

$$\tilde{P}_o(10,2,.10) = 0.16101, \quad \tilde{P}_1(10,2,.10) = 0.48050, \text{ etc. (15).}$$

and for the Poisson approximation (with $\lambda = 1.21306$)

$$P(X = 0|1.21306) = 0.29729, \quad P(x = 1|1.21306) = 0.26063, \text{ etc.}$$

The last two columns give the mean and variance of the various exact and approximating distributions. Note, also, that for the Poisson distribution the sum of the probabilities in each row may be less than one, reflecting the fact that a Poisson random variable assumes all non-negative integral values with positive probability.

In Tables II and V are given (a) the exact 2-step transition probabilities $P_{ij}^{(2)}$; b) the binomial approximation $\tilde{P}_{ij}^{(2)}$ given by (16); and (c) the binomial probabilities $b_{ij}^{(2)}$ as given by (19). In each of the tables all omitted entries are zero (to five decimal places) and the decimal point is omitted in the various probabilities given.

REFERENCES

[1] Ashby, E. 1935. The quantitative analysis of vegetation. Ann. Bot. Lond. 49:779-802.

[2] Barton, D. E., and David, F. N. 1959. A collector's problem. Trab. Estab. 10:75-88.

[3] ————, and ————. 1962. Combinatorial Chance. Charles Griffin, London.

[4] Blyth, C., and Curme, G. 1960. Estimation of a parameter in the classical occupancy problem. Biometrika 47:180-5.

[5] Chia, A. B. 1968. Random mating in a population of cyclic size. J. Appl. Prob. 5:21-30.

[6] Craig, C. C. 1953. On the utilization of marked specimens in estimating populations of flying insects. Biometrika 40:170-6.

[7] Curme, G., and Harkness, W. 1957. Probability distributions for air battles. Inst. Appl. Sci. No. 57, Univ. of Chicago.

[8] David, F. N. 1950. Two combinatorial tests of whether a sample has come from a given population. Biometrika 37:97-110.

[9] Dietz, K. 1966. On the model of Weiss for the spread of epidemics by carriers. J. Appl. Prob. 3:375-82.

[10] Downton, F. 1967. Epidemics with carriers: a note on a paper by Dietz. J. Appl. Prob. 4:264-70.

[11] Feller, W. 1968. An Introduction to Probability and Its Applications (3rd ed.). Wiley, New York.

[12] Khazanie, R. G., and McKean, H. E. 1966. A Mendelian Markov process with binomial transition probabilities. Biometrika 53:37-45.

[13] Mertz, D. B., and Davies, R. B. 1968. Cannibalism of the pupal stage by adult flour beetles: an experiment and a stochastic model. Biometrics 24:247-5.

[14] Nicholson, W. L. 1961. Occupancy probability distribution critical points. Biometrika 48:175-80.

[15] Patil, G. P., and Joshi, S. W. 1968. A Dictionary and Bibliography of Discrete Distributions. Oliver and Boyd, Edinburgh.

[16] Peto, S. 1953. A dose-response equation for the invasion of micro-organisms. Biometrics 9:320-35.

[17] Raff, M. S. 1956. On approximating the point binomial. J. Amer. Stat. Assoc. 51:293-303.

[18] Rashevsky, N. 1955. Note on a combinatorial problem in mathematical biology. Bull. Math. Biophys. 17:45-50.

[19] Sprott, D. A. 1957. Probability distributions associated with direct hits on targets. Bull. Math. Biophys. 19:163-70.

[20] Thomas, C. J. 1956. The estimation of bombers surviving an air battle. A.F.O.O.A. U-1287, T.M. 49.

[21] Thompson, H. R. 1958. The statistical study of plant distribution patterns using a grid of quadrats. Austral. J. Bot. 6:322-43.

[22] Thompson, M. 1951. Some tests for randomness in plant populations. Biometrika 38:102-11.

[23] Weiss, G. H. 1965. On the spread of epidemic carriers. Biometrics 21:481-90.

[24] Weiss, I. 1958. Limiting distributions in some occupancy problems. Ann. Math. Statist. 29:878-84.

Table 1. Power of A = 1 P - 0.10 R = 10
Exact CLOD and Approximating Binomial and Poisson Distributions are given below

n/x	0	1	2	3	4	5	6	7	8	9	10	Exp	Var
1	65132	34868										34868	22710
	65132	34868										34868	22710
	69220	25465	04684	00574	00053	00004						36788	36788
2	15120	50012	34868									1.19747	46089
	16101	48050	35849									1.19747	48050
	29720	36063	21873	08844	02682	00651	00132	00023	00003			1.21306	1.21306
3	01874	17378	45880	34868								2.13741	57855
	02377	17671	43786	36166								2.13741	61457
	11653	25050	26923	19291	10367	04457	01597	00490	00132	00031	00007	2.14959	2.14959
4	00145	02875	18150	43962	34868							3.10532	64716
	00250	03475	18091	41861	36323							3.10532	69457
	04437	13822	21529	22356	17411	10848	05632	02506	00976	00338	00105	3.11520	3.11520
5	00008	00279	03474	18517	42855	34868						4.08536	69182
	00020	00457	04087	18253	40765	36417						4.08536	74732
	01668	06827	13975	19069	19516	15978	10901	06375	03262	01484	00607	4.09365	4.09365
6		00017	00386	03869	18724	42135	34868					5.07176	72313
	00001	00045	00614	04473	18329	40059	36479					5.07176	78463
	00623	03163	08031	13596	17264	17536	14844	10770	06837	03859	01960	5.07889	5.07889
7		00001	00027	00472	04148	18855	41630	34868				6.06190	74629
		00004	00068	00733	04738	18369	39565	36524				6.06190	81238
	00232	01405	04263	08623	13081	15876	16056	13919	10558	07118	04320	6.06815	6.06815
8			00001	00036	00540	04356	18944	41255	34868			7.05442	76410
			00006	00089	00826	04930	18391	39201	36557			7.05442	83382
	00086	00606	02140	05037	08890	12552	14770	14896	13146	10312	07280	7.05998	7.05998
9				00002	00044	00596	04515	19008	40967	34868		8.04855	77822
				00008	00106	00900	05076	18404	38921	36583		8.04855	85087
	00032	00256	01031	02768	05573	08977	12050	13863	13956	12489	10058	8.05355	8.05355
10					00002	00052	00642	04642	19056	40738	34868	9.04382	78968
				00001	00011	00122	00960	05191	18412	38700	36603	9.04382	86475
	00012	00106	00481	01452	03284	05943	08963	11585	13103	13174	11920	9.04837	9.04837

Table 2. Power of A = 2P = 0.10 R = 10
Exact CLOD and the Two Binomial Approximations are given below

n/x	0	1	2	3	4	5	6	7	8	9	10	Exp	Var
1	87842	12158										12158	10680
	87842	12158										12158	10680
	98813	01187										01187	01173
2	52966	34876	12158									59191	48471
	53169	33979	12851									59682	49765
	56200	37533	06267									50067	37533
3	20784	35064	31995	12158								1.35526	88788
	21796	33591	31533	13080								1.35896	92763
	17445	41328	32636	08591								1.32372	73964
4	05637	18722	32827	30657	12158							2.24977	1.14318
	06512	18564	31386	30345	13194							2.25144	1.21410
	03761	19118	36441	30872	09808							2.23846	98578
5	01126	06382	18697	31752	29885	12158						3.19361	1.29128
	01520	06932	18320	30313	29653	13262						3.19434	1.38649
	00624	05492	19337	34039	29960	10548						3.18862	1.15516
6	00173	01533	06907	18732	31114	29383	12158					4.15861	1.38424
	00290	01938	07355	18234	29674	29201	13307					4.15897	1.49644
	00084	01140	06421	19295	32614	29402	11044					4.15598	1.27728
7	00021	00275	01804	07255	18768	30689	29031	12158				5.13452	1.44799
	00047	00431	02221	07639	18191	29248	28884	13340				5.13472	1.57235
	00010	00185	01524	06984	19201	31672	29025	11400				5.13296	1.36906
8	00002	00038	00351	01997	07502	18798	30385	28770	12158			6.11686	1.49457
	00007	00080	00534	02420	07840	18166	28942	28648	13364			6.11698	1.62807
	00001	00025	00279	01813	07356	19103	31003	28753	11666			6.11585	1.44040
9		00004	00053	00410	02142	07686	18822	30156	28568	12158		7.10333	1.53015
	00001	00013	00106	00611	02568	07990	18149	28712	28467	13383		7.10341	1.67079
		00003	00041	00362	02035	07619	19014	30504	28548	11874		7.10264	1.49736
10			00006	00065	00458	02255	07830	18842	29978	28409	12158	8.09263	1.55825
		00002	00018	00128	00673	02683	08107	18136	28533	28323	13398	8.09269	1.70463
			00005	00059	00434	02210	07812	18935	30117	28387	12040	8.09214	1.54387

Table 3. Power of A = 1 P - 0.50 R = 20
Exact CLOD and Approximating Binomial and Poisson Distributions are given below

n/x	0	1	2	3	4	5	6	7	8	9	10	Exp	Var
1	1.00000												
	1.00000												
	99995	00005										00005	00005
2	99366	00634										00634	00630
	99367	00632	00001									00634	00632
	98661	01330	00009									01348	01348
3	92265	07645	00090									07825	07393
	92377	07422	00199	00002								07825	07621
	89851	09616	00515	00018								10702	10702
4	74186	23977	01804	00033								27683	23825
	75060	22324	02490	00123	00002							27683	25768
	72012	23644	03882	00425	00035	00002						32834	32834
5	49961	40051	09244	00726	00018							60788	46895
	52302	36194	10019	01387	00096	00003						60788	53398
	50830	34396	11637	02625	00444	00060	00007	00001				67668	67668
6	27934	44321	22682	04661	00390	00012						1.05288	72684
	31420	40122	21348	06058	00967	00082	00003					1.05288	86812
	32198	36489	20676	07810	02213	00502	00095	00015	00002			1.13325	1.13325
7	12967	35718	33987	14272	02799	00249	00009					1.59003	98632
	16469	33883	29876	14634	04301	00758	00074	00003				1.59003	1.22886
	18683	31342	26289	14700	06165	02068	00578	00139	00029	00005	00001	1.67756	1.67756
8	05008	22037	35015	26048	09816	01890	00178	00007				2.20047	1.23409
	07628	23154	30749	23333	11067	03359	00637	00069	00003			2.20047	1.59521
	10106	23163	26546	20281	11621	05327	02035	00666	00191	00049	00011	2.29204	2.29204
9	01613	10719	26502	31904	20500	07233	01386	00137	00006			2.86927	1.46440
	03158	13302	24903	27195	19091	08935	02788	00559	00065	00003		2.86927	1.95452
	05168	15311	22681	22399	16591	09831	04854	02055	00761	00250	00074	2.96274	2.96274
10	00433	04185	15339	28115	28492	16621	05620	01078	00111	00005		3.58486	1.67554
	01180	06597	16588	24720	24174	16210	07549	02410	00505	00063	00004	3.58486	2.29974
	02525	09290	17088	20955	19272	14180	08694	04569	02101	00859	00316	3.67879	3.67879

Table 4. Power of A = 1 P = -0.10 R = -10
Exact CLOD and Approximating Binomial and Poisson Distributions are given below

n/x	0	1	2	3	4	5	6	7	8	9	10	Exp	Var
1	61446	38554										38554	23690
	61446	38554										38554	23690
	69220	25465	04684	00574	00053	00004						36788	36788
2	15772	45674	38554									1.22783	49136
	14906	47405	37689									1.22783	47405
	29729	36063	21873	08844	02682	00651	00132	00023	00003			1.21306	1.21306
3	02653	17118	41675	38554								2.16131	63682
	02185	16892	43530	37393								2.16131	60422
	11653	25050	26923	19291	10367	04457	01597	00490	00132	00031	00007	2.14959	2.14959
4	00345	03799	17441	39860	38554							3.12479	72742
	00229	03273	17530	41725	37243							3.12479	68371
	04437	13822	21529	22356	17411	10848	05632	02506	00976	00338	00105	3.11520	3.11520
5	00038	00615	04417	17551	38825	38554						4.10174	78874
	00019	00427	03902	17818	40681	37153						4.10174	73688
	01668	06827	13975	19069	19516	15978	10901	06375	03262	01484	00607	4.09365	4.09365
6	00004	00080	00812	04800	17594	38156	38554					5.08587	83287
	00001	00042	00581	04308	17975	40002	37092					5.08587	77486
	00623	03163	08031	13596	17264	17536	14844	10770	06837	03859	01960	5.07889	5.07889
7		00009	00119	00958	05059	17612	37689	38554				6.07428	86610
		00003	00064	00699	04590	18070	39524	37049				6.07428	80330
	00232	01405	04263	08623	13081	15876	16056	13919	10558	07118	04320	6.06815	6.06815
8		00001	00015	00152	01071	05245	17618	37343	38554			7.06545	89201
			00006	00084	00793	04797	18134	39170	37017			7.06545	82538
	00086	00606	02140	05037	08890	12552	14770	14896	13146	10312	07280	7.05998	7.05998
9			00002	00020	00181	01159	05386	17620	37078	38554		8.05849	91277
				00008	00101	00868	04956	18178	38897	36991		8.05849	84302
	00032	00256	01031	02768	05573	08977	12050	13863	13956	12489	10058	8.05355	8.05355
10				00002	00026	00205	01231	05496	17618	36838	38554	9.05287	92977
				00001	00010	00117	00930	05081	18211	38680	36971	9.05287	85743
	00012	00106	00481	01452	03284	05943	08963	11585	13103	13174	11920	9.04837	9.04837

Table 5. Power of A = 2 P = -0.10 R = -10
Exact CLOD and the Two Binomial Approximations are given below

n/x	0	1	2	3	4	5	6	7	8	9	10	Exp	Var
1	85136	14864										14864	12655
	85136	14864										14864	12655
	96158	03842										03842	03694
2	49917	35219	14864									64947	52494
	49653	36143	14205									64552	51292
	51756	40371	07873									56117	40371
3	20767	32234	32135	14864								1.41097	95469
	19870	33464	32683	13982								1.40778	91854
	15894	40344	34135	09627								1.37495	74479
4	06621	17718	30061	30736	14864							2.29504	1.25691
	05850	17839	31298	31142	13870							2.29343	1.19255
	03415	18118	36060	31863	10564							2.28042	98034
5	01726	06971	17492	29009	29938	14864						3.23054	1.44992
	01352	06515	17808	30257	30265	13803						3.22977	1.36237
	00566	05135	18638	33826	30695	11141						3.22372	1.14524
6	00385	02157	07380	17411	28381	29422	14864					4.18969	1.57781
	00257	01794	07021	17836	29635	29699	13758					4.18930	1.47322
	00076	01057	06100	18768	32483	29984	11532					4.18602	1.26556
7	00076	00556	02457	07655	17370	27961	29061	14864				5.16137	1.66803
	00041	00395	02092	07358	17865	29219	29304	13726				5.16114	1.55085
	00009	00171	01435	06702	18682	31583	29506	11813				5.15917	1.35674
8	00013	00124	00681	02668	07851	17344	27659	28795	14864			6.14050	1.73510
	00006	00072	00497	02303	07598	17890	28920	29011	13702			6.14036	1.60831
	00001	00023	00261	01724	07107	18757	30940	29162	12026			6.13908	1.42804
9	00002	00025	00163	00775	02826	07997	17326	27432	28590	14864		7.12446	1.78700
	00001	00011	00098	00576	02462	07777	17909	28695	28787	13684		7.12437	1.65262
		00003	00039	00342	01949	07397	18720	30456	28904	12191		7.12348	1.48526
10		00004	00034	00194	00850	02947	08110	17313	27255	28428	14864	8.11173	1.82839
		00002	00016	00119	00639	02586	07916	17925	28519	28609	13669	8.11166	1.68787
			00005	00055	00413	02128	07613	18680	30080	28702	12324	8.11103	1.53215

Table 6. Power of A = 1 P = -0.50 R = -20
Exact CLOD and Approximating Binomial and Poisson Distributions are given below

n/x	0	1	2	3	4	5	6	7	8	9	10	Exp	Var
1	99970	00030										00030	00030
	99970	00030										00030	00030
	99995	00005										00005	00005
2	97724	02246	00030									02306	02313
	97707	02279	00013									02306	02279
	98661	01330	00009									01348	01348
3	87175	11934	00861	00030								13746	13759
	86874	12515	00601	00010								13746	13116
	89851	09616	00515	00018								10702	10702
4	68330	26033	05041	00565	00030							37932	37379
	67130	28132	04421	00309	00008							37932	34335
	72012	23644	03882	00425	00035	00002						32834	32834
5	47068	35699	13584	03172	00447	00030						74322	71253
	44726	39045	13634	02380	00208	00007						74322	63274
	50830	34396	11637	02625	00444	00060	00007	00001				67668	67688
6	28859	36369	22888	09089	02378	00386	00030					1.21034	1.12004
	25877	39235	24787	08351	01583	00160	00007					1.21034	96619
	32198	36489	20676	07810	02213	00502	00095	00015	00002			1.13325	1.13325
7	15981	29958	28010	16793	06926	01954	00348	00030				1.76130	1.56589
	13149	30945	31212	17489	05880	01186	00133	00006				1.76130	1.31813
	18683	31342	26289	14700	06165	02068	00578	00139	00029	00005	00001	1.67756	1.67756
8	08098	20976	27102	22687	13396	05693	01696	00323	00030			2.37964	2.02748
	05935	20102	29788	25224	13350	04522	00957	00116	00006			2.37964	1.67180
	10106	23163	26546	20281	11621	05327	02035	00666	00191	00049	00011	2.29204	2.29204
9	03797	12899	21869	24195	19188	11283	04910	01523	00305	00030		3.05225	2.48952
	02404	11105	22794	27294	21010	10782	03789	00811	00104	00006		3.05225	2.01711
	05168	15311	22681	22399	16591	09831	04854	02055	00761	00250	00074	2.96274	2.96274
10	01663	07127	15257	21423	21801	16771	09863	04375	01400	00292	00030	3.76889	2.94225
	00882	05337	14527	23430	24801	18001	09073	03136	00711	00096	00006	3.76889	2.34844
	02525	09290	17088	20955	19272	14180	08694	04569	02101	00859	00316	3.67879	3.67879

ESTIMATING THE BERNOULLI PARAMETER AND THE AVERAGE SAMPLE SIZE FROM TRUNCATED SAMPLES

A. CLIFFORD COHEN
The University of Georgia

SUMMARY

This paper is concerned with maximum likelihood estimation of the Bernoulli parameter p in a dichotomous population based on truncated samples employed in testing the hypothesis H_o: $p \leq p_o$ against the alternative H_a: $p > p_o$. Sample specimens are examined one at a time in sequence until either the kth non-conforming (infected or defective) specimen is found or until the Kth conforming (noninfected) specimen is found. If the kth nonconforming specimen is found prior to observing the Kth conforming specimen, sampling is terminated with rejection of H_o. Otherwise sampling is terminated with acceptance of H_o on finding the Kth conforming specimen. The maximum likelihood estimate $\hat{p}$ based on results of a sequence of m tests is shown to be merely the ratio of the total number of nonconforming specimens found to the total number of specimens observed. The asymptotic variance of this estimator is shown to be approximately $\hat{p}(1-\hat{p})/\Sigma_1^m y_i$, where $\Sigma_1^m y_i$ is the total number of specimens observed. Formulas are given for determining the expected sample sizes, and illustrative examples are included to provide comparisons for assessing the reduction in sample sizes to be achieved through truncation. The principal results contained here were previously given in reference [2].

I. INTRODUCTION

The need to estimate the Bernoulli parameter, p, and to test hypotheses concerning this parameter occurs frequently in the life sciences. For example, it may become necessary, in an investigation of parasite infestation in certain animal populations, to test hypotheses of the form H_o: $p \leq p_o$ against alternatives of the form H_a: $p > p_o$, where p is the proportion of the members of a population that are infected. Based on the results on one or more such tests, it may also be necessary to estimate p. In a typical test, the number of infected specimens in a random sample of size n is determined. If this number is equal to or greater than a rejection number k, H_o is rejected in favor of H_a. Otherwise H_o is accepted. The maximum likelihood estimate of $\hat{p}$ in this case is the ratio of the number of "infected" specimens to the total sample size, n. In the design of such tests, n and k are determined such that specified values of α and β, the respective probabilities of type I and type II error, will be attained. Single sample tests of this type are widely used in connection with product acceptance sampling. Several excellent tables of such plans together with their operating characteristic curves are available. One of the best known and perhaps most comprehensive sets of these tables is due to Dodge and Romig [5].

In order to conserve testing effort and test materials, fixed-size sampling plans are sometimes modified to the extent that sampling is terminated (i.e., curtailed) when it becomes possible to reach a decision with respect to acceptance or rejection of a given lot of product (i.e., acceptance or rejection of a given hypothesis) prior to inspection of the entire sample. Such curtailed or truncated plans have recently been investigated by Burr [1], Phatak and Bhatt [9], Craig [3, 4] and the writer [2]. In each of these investigations, the writers have been primarily concerned with product acceptance sampling. However, their results are equally applicable to hypothesis testing and to parameter estimation in the field of Biometrics.

2. THE HYPOTHESIS AND THE TEST

The hypothesis to be tested is H_o: $p \leq p_o$, and the alternative is H_a: $p > p_o$. The test is specified by an acceptance number K and a rejection number k. For convenience, we will assume that we are dealing with infected versus noninfected members of some specified population of interest. Randomly selected individual members of this population are examined in sequence until either

1. An accumulated total of K "noninfected" members have been found, in which case H_o is accepted and sampling is terminated or until
2. An accumulated total of k "infected" members have been found, in which case sampling is terminated with rejection of H_o in favor of H_a.

The number of items, Y, examined in order to reach a decision with respect to acceptance or rejection of the hypothesis being tested is thereby a discrete random variable which may assume the values k, k+1, ..., n, where

$$n = k + K - 1. \tag{1}$$

Sampling may thus be terminated after examining as few as k individual members, and in no case will more than n members be inspected for a given test.

3. THE RANDOM VARIABLE, Y, NUMBER OF SPECIMENS INSPECTED

Let p designate the proportion of population members in which the characteristic of interest is present. We are concerned with random sampling and with populations that are infinite or at least large enough to justify the assumptions that p remains constant from trial to trial and that the trials are stochastically independent. The probability function of Y may be expressed as

$$f(y;p) = f(y \cap A;\ p) + f(y \cap R;\ p), \tag{2}$$

where $f(y \cap A;\ p)$ is the joint probability that Y = y and that H_o will be accepted, whereas $f(y \cap R;\ p)$ is the joint probability that Y = y and that H_o will be rejected. It follows that

$$f(y \cap R;\ p) = \binom{y-1}{k-1} p^k q^{y-k};\ y = k, k+1, \ldots, n, \tag{3}$$

and

$$f(y \cap A;\ p) = \binom{y-1}{K-1} q^K p^{y-K};\ y = K,\ K+1,\ \ldots,\ n, \tag{4}$$

where $q = 1 - p$.

Functions (3) and (4) are recognized as probability functions of one-parameter negative binomial distributions. The probability function (2) may be expressed as

$$f(y;\ p) = \begin{cases} f(y \cap R;\ p);\ y = k,\ k+1,\ \ldots,\ K-1. \\ f(y \cap R;\ p) + f(y \cap A;\ p);\ y = K,\ K+1,\ \ldots,\ n. \end{cases} \tag{5}$$

On substituting (3) and (4) into (5), we have

$$f(y;\ p) = \begin{cases} \binom{y-1}{k-1} p^k q^{y-k};\ y = k,\ k+1,\ \ldots,\ K-1, \\ \binom{y-1}{k-1} p^k q^{y-k} + \binom{y-1}{K-1} q^K p^{y-K};\ y = K,\ K+1,\ \ldots,\ n, \\ 0,\ \text{elsewhere}. \end{cases} \tag{6}$$

The probability of rejecting H_o follows from (3) when we sum on y from k to n. Thus

$$P(R) = \sum_{y=k}^{n} \binom{y-1}{k-1} p^k q^{y-k}\ . \tag{7}$$

It has been shown by Patil [7, 8], by Morris [6], and perhaps by others that the following identity holds between the binomial and the negative binomial distributions:

$$\sum_{z=k}^{n} \binom{n}{z} p^z q^{n-z} = \sum_{y=k}^{n} \binom{y-1}{k-1} p^k q^{y-k}\ . \tag{8}$$

Therefore, the probability of rejecting H_o can be expressed in the alternate form

$$P(R) = \sum_{z=k}^{n} \binom{n}{z} p^z q^{n-z}\ , \tag{9}$$

where Z is the number of infected specimens present in a sample of size n and where n is given by (1).

It follows, consequently, that the operating characteristic

curve of the truncated sampling plan described above as specified by k and K is identical with the operating characteristic curve of an ordinary single sampling plan specified by n and k, where $n = k + K - 1$. Of course, truncation permits a decision to be reached with fewer inspections than when fixed size samples are employed.

4. ESTIMATION USING TRUNCATED SAMPLES

Suppose that m tests of hypotheses have been made in accordance with the procedure described in Section 2. Let a be the number of tests that resulted in acceptance of H_o, and let r be the number that resulted in rejection so that

$$m = a + r \; . \tag{10}$$

Let the number of infected specimens found and the number of items inspected be recorded for each test. The sample data then consists of the paired values (z_1,y_1), (z_2,y_2), ..., (z_a,y_a), (k,y_{a+1}) ..., (k,y_{a+r}), where z_i $(i = 1, ..., a)$ is the number of infected specimens found in the ith test which resulted in acceptance $(z_i < k)$, and k, of course, is the number of infected specimens found in each test which resulted in rejection of H_o. We could describe our sample more concisely as consisting of the paired values (z_i,y_i); $i = 1, 2, ..., m$, with $z_i < k$ for tests resulting in acceptance and $z_i = k$ for tests resulting in rejection. Using (6), the likelihood function for such a sample follows as

$$L\,[(z_1,y_1),\,(z_2,y_2),\,...,\,(z_m,y_m)] = \tag{11}$$

$$\prod_{i=1}^{r} \binom{y_i - 1}{k-1} p^k q^{y_i - k} \quad \prod_{j=1}^{a} \binom{y_j - 1}{K-1} q^K p^{y_j - K} \; .$$

On taking logarithms of (11), differentiating with respect to p, equating to zero, and solving for p we obtain

$$\hat{p} = \frac{\sum_{j=1}^{a} y_j - aK \quad + rk}{\sum_{i=1}^{m} y_i} \; . \tag{12}$$

We note that in (12),

rk = number of infected specimens in r tests in which H_o was rejected;

aK = number of non-infected specimens in a tests in which H_o was accepted;

$\sum_1^a y_j$ = total number inspections from a tests in which H_o was accepted;

$\sum_1^m y_i$ = total number inspections from $m = a + r$ tests.

Accordingly, (12) can be written as

$$\hat{p} = \frac{\text{Total number of infected specimens}}{\text{Total number of specimens inspected}} .$$

The asymptotic variance of $\hat{p}$ in this case can be obtained as $V(\hat{p}) \sim -1/E(\partial^2 \ln L/\partial p^2)$, which after certain algebraic reduction becomes

$$V(\hat{p}) \sim \hat{p}\hat{q}/mE(Y). \tag{14}$$

The expected value of Y is given in Section 5, which follows. However, for a sufficiently large number of tests, the mean of the observed values of Y should provide a reasonable approximation to E(Y). Accordingly, the variance of (14) might be approximated as

$$V(\hat{p}) \sim \hat{p}\hat{q}/\Sigma_1^m y_i . \tag{15}$$

5. AVERAGE SAMPLE SIZE IN TRUNCATED SAMPLES

In the notation of this paper, the average sample size (or average sample number, ASN) is merely the expected value of Y:

$$ASN = E(Y) . \tag{16}$$

It follows from (6) that

$$E(Y) = \sum_{y=k}^{n} y \binom{y-1}{k-1} p^k q^{y-k} + \sum_{y=K}^{n} y \binom{y-1}{K-1} q^K p^{y-K} , \tag{17}$$

which, using the identity of (8), becomes

$$E(Y) = \frac{k}{p} \{1 - B[p,n+1,k]\} + \frac{K}{q} B[p,n+1,k-1] , \tag{18}$$

where

$$B[p,n,k] = \sum_{x=0}^{k} \binom{n}{x} p^x q^{n-x}, \text{ with } n = k + K - 1. \tag{19}$$

The expression for E(Y) in (18), above, was given by Phatak and Bhatt [9]. The following alternate form, which is perhaps a bit simpler for use in making calculations, follows from (18) after certain algebraic manipulations:

$$E(Y) = \frac{k}{p} P(R) + \frac{K}{q} [P(A) - b(k-1; n,p)], \tag{20}$$

where

$$b(k; n,p) = \binom{n}{k} p^k q^{n-k}, \text{ and} \tag{21}$$

$$P(R) = \sum_{z=k}^{n} \binom{n}{z} p^z q^{n-z} = 1 - B[p,n,k-1] \tag{22}$$

is the probability of rejecting H_o. Of course P(A) = 1 - P(R) is the probability of accepting H_o.

The following tabulations provide the readers with an indication of the reduction in sample sizes to be achieved through truncation in selected tests.

n = 10, k = 4, K = 7			n = 20, k = 5, K = 16		
p	P(R)	ASN	p	P(R)	ASN
0	.00000	7.00	0	.00000	16.00
.10	.01280	7.74	.10	.04317	17.57
.15	.04997	8.09	.15	.17015	17.86
.20	.12087	8.35	.20	.37035	17.49
.25	.22412	8.49	.25	.58516	16.51
.30	.35039	8.50	.30	.76249	15.16
.35	.48617	8.37	.35	.88180	13.69
.40	.61772	8.13	.40	.94905	12.29
.45	.73396	7.79	.45	.98114	11.05
.50	.82813	7.39	.50	.99409	9.98
1.00	1.00000	4.00	1.00	1.00000	5.00

REFERENCES

[1] Burr, I. W. 1957. "Average sample number under curtailed or truncated sampling," Industrial Quality Control, 13, No. 8, pp. 5-7.

[2] Cohen, A. Clifford, 1970. "Curtailed attribute sampling," Technometrics, 12:295-8.

[3] Craig, C. C. 1968. "The ASN for truncated single and double attribute acceptance sampling plans," ASOC Convention Transactions, pp. 63-7.

[4] ———. 1968. "The average sample number for truncated single and double attribute sampling plans," Technometrics 10:685-92.

[5] Dodge, Harold F., and Romig, Harry G. 1959. Sampling Inspection Tables 2nd. Ed., John Wiley and Sons, Inc.

[6] Morris, K. W. 1963. "A note on direct and inverse binomial sampling," Biometrika, 50:544-5.

[7] Patil, G. P. 1960. "On the evaluation of the negative binomial distribution with examples," Technometrics, 2:501-5.

[8] ———. 1963. "Note on the equivalence of the binomial and inverse binomial acceptance sampling plans and an acknowledgement," Technometrics, 5:119-21.

[9] Phatak, A. G., and Bhatt, M. M. 1967. "Estimation of fraction defective in curtailed sampling plans," Technometrics, 9:219-28.

ON ACCEPTANCE SAMPLING WITHOUT REPLACEMENT

K. G. JANARDAN
G. P. PATIL
The Pennsylvania State University

SUMMARY

The authors discuss hypergeometric and inverse hypergeometric acceptance plans and show that these plans are equivalent and that, when properly modified, they are in fact identical. The authors also discuss certain associated problems of estimation.

1. INTRODUCTION

Acceptance sampling plans based on sampling with replacement have received considerable attention in statistical literature. While hypergeometric and inverse hypergeometric probability distributions have appeared in varied publications, acceptance sampling plans based on sampling without replacement, as such, do not seem to have been discussed. The purpose of this paper is to discuss hypergeometric and inverse hypergeometric acceptance sampling plans and also to show that these plans are equivalent and that, when properly modified, they are in fact identical. The paper also discusses certain associated problems of estimation.

2. DESCRIPTION OF PLANS

Plan 1: Hypergeometric Acceptance Sampling Plan (HAS Plan)

Inspect a sample of size n drawn at random without replacement from a lot of N items containing k defectives. Accept the lot if the number of defectives are less than d. Reject the lot if the number of defectives are d or more.

Following Cohen [1] we shall introduce the following notation: let x denote the number of defectives found. Let $f(x\cap R,p)$ be the joint probability that the number of defectives equals x and the lot is accepted, and $f(x\cap R,p)$ be the joint probability that the number of defectives equals x and the lot is rejected. Then the probability function (pf) associated with this plan would be

$$f(x;p) = f(x\cap A;p) + f(x\cap R;p),$$

where

$$f(x\cap A;p) = \frac{\binom{n}{x}\binom{N-n}{k-x}}{\binom{N}{k}} \quad \text{for } x = 0, 1, 2, \ldots, d-1 \tag{2.1}$$

$$f(x\cap R;p) = \frac{\binom{n}{x}\binom{N-n}{k-x}}{\binom{N}{k}} \quad \text{for } x = d, d+1, \ldots, n \tag{2.2}$$

and clearly the probability of rejecting a lot is $H(d;n,N,k)$, where

$$H(d;n,N,k) = \sum_{x=d}^{\min(n,k)} \frac{\binom{n}{x}\binom{N-n}{k-x}}{\binom{N}{k}}. \tag{2.3}$$

The HAS Plan is specified by producer's risk α at the Acceptable Quality Level (AQL) $\frac{100\ k_o}{N}$ and consumer's risk β at the Lot Tolerance Percent Defective (LTPD) $\frac{100\ k_1}{N}$. These specifications determine the choice of the sample size n and the rejection number d of the plan and are given by the following:

$$H(d;n,N,k_o) = \alpha \tag{2.4}$$

$$H(d;n,N,k_1) = 1-\beta \tag{2.5}$$

Plan 2: Inverse Hypergeometric Acceptance Sampling Plan (IHAS Plan)

One at a time without replacement, inspect items selected randomly from the lot of N items containing k defectives until either δ defectives have been observed or until η units have been inspected. Reject the lot if δ defectives are observed. Accept the lot if n items are inspected, provided that the number of defectives found is less than δ. Let z denote the number of defectives when a lot is accepted or the number of items inspected when a lot is rejected. Then the pf associated with this plan is

$$f(z;p) = f(x\cap A;p) + f(y\cap R;p),$$

where $f(x\cap A;p)$ is as defined earlier in (2.1) and

$$f(y\cap R;p) = \frac{\binom{y-1}{\delta-1}\binom{N-y}{k-\delta}}{\binom{N}{k}} \quad y = \delta,\ \delta+1,\ldots,n \tag{2.6}$$

and, clearly, the probability of rejecting a lot is $1H(\eta;\delta,N,k)$, where

$$1H(\eta;\delta,N,k) = \sum_{y=\delta}^{\eta} \frac{\binom{y-1}{\delta-1}\binom{N-y}{k-\delta}}{\binom{N}{k}} \tag{2.7}$$

The specifications of IHAS Plan are again producer's risk α at the AQL $\frac{100\ k_o}{N}$, and the Consumer's risk β at the LTPD $\frac{100\ k_1}{N}$. These specifications determine η and δ by the following two equations:

$$1H(\eta;\delta,N,k_o) = \alpha \tag{2.8}$$

$$1H(\eta;\delta,N,k_1) = 1-\beta \tag{2.9}$$

Plan 3: Modified Hypergeometric Acceptance Sampling Plan (MHAS Plan)

Inspect items selected from the lot one at a time and without replacement until either

i) an accumulated total of d defectives is found, in which case, reject the lot, or until

ii) an accumulated total of c (= n-d+1) nondefectives is found in which case, reject the lot. Clearly the number of items inspected y is a random variable, and the pf is given by

$$f(y;p) = f(y \cap A;p) + f(y \cap R;p),$$

where $f(y \cap A;p)$ is as defined earlier in (2.6), and

$$f(y \cap A;p) = \frac{\binom{y-1}{c-1}\binom{N-y}{N-k-c}}{\binom{N}{N-k}} \qquad (2.10)$$

for $y = c, c+1, \ldots, n$.

3. INTERRELATIONS OF THE PLANS

Patil [4] has shown the equivalence between the Binomial and Inverse Binomial acceptance sampling plans. We shall now show here the equivalence of the HAS and IHAS plans.

3.1 Equivalence of HAS and IHAS Plans:

The HAS and IHAS Plans are clearly quite different in approach and character. However, by the identity between the right and left tails of the hypergeometric and inverse hypergeometric distributions provided by

$$H(d;n,N,k) = IH(\eta;\delta,N,k), \qquad (3.1)$$

we get the relationship between the parameters of these two plans as $\delta = d$ and $\eta = n$.

A probabilistic proof for (3.1) may be written down in brief, as follows:

$$H(d;n,N,k) = \frac{\sum_{x=d}^{n}\binom{n}{x}\binom{N-n}{k-x}}{\binom{N}{k}}$$

= Probability that in n items drawn d or more defectives are found.

= Probability that, at most, n items are required to be drawn for d defectives.

$$= \sum_{y=d}^{n} \frac{\binom{y-1}{d-1}\binom{N-y}{k-d}}{\binom{N}{k}}$$

$$= IH(n;d,N,k)$$

as obtained in Lieberman and Owen [2] in a slightly different form.

A mathematical proof for (3.1) independent of the present probability argument should be an interesting problem to work out.

The identity (3.1) reveals that the operating characteristic curve (oc-curve) of the MHAS plan, as specified by d and c, is identical with the oc-curve of HAS plan specified by n and d when n = d + c-1. But the curtailment permits a decision to be reached with fewer inspections than when fixed-size samples are used.

3.2 The Identity Between Modified HAS and IHAS Plans:

Obvious curtailments are possible for both the HAS and the IHAS plans. The modified HAS plan is discussed in section 2. The average sample number (ASN) required in this plan is obtained as follows:

$$\text{ASN(MHAS)} = \frac{\sum_{y=d}^{n} y \binom{y-1}{d-1}\binom{N-y}{k-d}}{\binom{N}{k}} + \frac{\sum_{y=c}^{n} y \binom{y-1}{c-1}\binom{N-n}{N-k-c}}{\binom{N}{k}},$$

where c = n-d + 1.

$$= \frac{d(N+1)}{k+1} \frac{\sum_{y=d}^{n} \binom{y}{d}\binom{N-y}{k-d}}{\binom{N+1}{k+1}} + \frac{c(N+1)}{N-k+1} \frac{\sum_{y=c}^{n} \binom{y}{c}\binom{N-y}{N-k-c}}{\binom{N+1}{N-k+1}}$$

$$= \frac{d(N+1)}{k+1} IH(n+1;\ d+1,\ N+1,\ k+1) +$$

$$\frac{c(N+1)}{N-k+1} IH(n+1;\ c+1,\ N+1,\ N-k+1)$$

$$= \frac{d(N+1)}{k+1} H(d+1;\ n+1,\ N+1,\ k+1) +$$

$$\frac{(n-d+1)\ (N+1)}{N-k+1} H(n-d+2;\ n+1,\ N+1,\ N-k+1) \qquad (3.2)$$

On lines similar to the HAS plan, modifications to IHAS plan may be written as follows: a lot is inspected by selecting items at random, one at a time and without replacement, until either δ defectives are found or $\eta-\delta+1$ nondefectives are observed. The lot is accepted if there are $\eta-\delta+1$ nondefectives, rejected if there are δ defectives. Clearly, the sample size actually needed is a random variable, as before. The ASN for MIHAS can be similarly reduced to

$$\text{ASN(MIHAS)} = \frac{\delta(N+1)}{k+1} H(\delta+1;\ \eta+1,\ N+1,\ k+1) + \frac{(\eta-\delta+1)\ (N+1)}{N-k+1} H(\eta-\delta+2;\ \eta+1,\ N+1,\ N-k+1). \quad (3.3)$$

Since $\delta = d$ and $\eta = n$ (3.2) and (3.3) reveal that both MHAS and MIHAS plans have the same ASN. In fact, the comparison of the descriptions of MHAS and MIHAS plans bring out that both acceptance sampling plans are the same. Thus it is clear that, while HAS and IHAS plans are equivalent, MHAS and MIHAS plans are identical.

4. UNBIASED ESTIMATES OF THE FRACTION DEFECTIVES AND THE VARIANCES OF THE ESTIMATES UNDER THE THREE PLANS

The maximum likelihood estimation of the fraction of defectives in curtailed sampling plans in the case with replacement has been considered by Phatak and Bhatt [5]. The maximum likelihood estimation of the fraction defective in curtailed acceptance sampling plans without replacement is rather more involved. Hence, we have considered here only the unbiased estimation.

Suppose that m lots have been subjected to inspection under one of the plans described in Section 2. Let 'a' denote the total number of lots that were accepted and 'r' denote the number of rejected lots so that $m = a+r$. Let the number of defectives found and the number of items inspected be recorded for each lot. Let x_i denote the number of defectives found in the sample from the ith lot. Let y_i denote the number of items inspected in the ith lot. The sample data then consists of the paired values (x_i, y_i) $i = 1,2,\ldots,m$. It is well known (Sandelius [6]) that if x defectives are obtained in a direct hypergeometric sample of size n, then $p' = \frac{x}{n}$ is an unbiased estimate of $P = \frac{k}{n}$ and the estimated variance of the estimate is $V^*(p') = \frac{N-n}{n} \frac{p'(1-p')}{n-1}$ and in inverse hypergeometric sampling $p'' = \frac{d-1}{n-1}$ $(d>2)$ is an unbiased estimate of $P = \frac{k}{n}$ and the estimated variance of p" is $V^*(p") = \frac{p"(1-p")}{n-2} \frac{N-n+1}{N}$.

The sample data, the unbiased estimate, variance of the estimate, and estimated variance of the estimate under the three plans discussed above are recorded in the table on the following page.

REFERENCES

[1] Cohen, A. C. 1969. Curtailed attribute sampling (Type Script) University of Georgia.

[2] Lieberman, G. J., and Owen, D. B. 1961. Tables of hypergeometric probability distribution. 3-8. Stanford University Press.

[3] Patil, G. P. 1960. On the evaluation of the negative binomial distribution with examples. Technometrics 2:501-5.

[4] ———. 1963. On the equivalence of the binomial and inverse binomial acceptance sampling plans and an acknowledgement. Technometrics 5:119-21.

[5] Phatak, A. G., and Bhatt, N. M. 1967. Estimation of the fraction defectives in curtailed sampling plans by attributes. Technometrics 9:123-7.

[6] Sandelius, M. 1951. Unbiased estimation based on inverse hypergeometric sampling. Kungl. Lantbrukshogskolans Annaler, 18:123-7.

Table of Unbiased Estimates, Variances of Estimates and Estimated Variances

Plan	Sample Data	Unbiased Estimate	Variance V of the estimate and the estimated Variance V* of the estimate.
HAS	(x_i, n) $i = 1,2,\ldots,m$	$\tilde{p}_1 = \frac{1}{m} \sum_{i=1}^{m} \frac{x_i}{n}$	$V_1 = \frac{1}{m} \frac{P(1-P)}{n} \frac{N-n}{N-1}$ $V_1^* = \frac{1}{m} \frac{\tilde{p}_1 (1-\tilde{p}_1)}{n} \frac{N-n}{N-1}$
IHAS	(x_i, y_i) $\left.\begin{matrix} x_i < d \\ y_i = n \end{matrix}\right\} i = 1,2,\ldots,a$ $x_i = d$, $i = a+1,\ldots,m$	$\tilde{p}_2 = \frac{1}{m}\{\sum_{i=1}^{a} \frac{x_i}{n} + \sum_{j=1}^{r} \frac{d-1}{y_i-1}\}$	$V_2 = \frac{1}{m^2} [E(A) \frac{P(1-P)}{n} \frac{N-n}{N-1} + (V(A) + V(R)) P^2 + E(R) V(\frac{d-1}{y-1})]$ Where A denotes the accepted lots, R, the rejected lots. $V_2^* = \frac{1}{m^2} [E(A) \frac{\tilde{p}_2(1-\tilde{p}_2)}{n} \frac{N-n}{N-1} + (V(A) + V(R)) p_2{}^2 + E(R) \frac{\tilde{p}_2(1-\tilde{p}_2)}{n-2} \frac{N-n+1}{N}]$
MHAS	(x_i, y_i) $x_i > d$ $i = 1,2,\ldots,a$ $x_i = d$ $i = a+1,\ldots,m$	$\tilde{p}_3 = \frac{1}{m} \{\sum_{i=1}^{r} \frac{d-1}{y_i-1} +$ $(a - \sum_{j=1}^{a} (\frac{c-1}{y_j-1})\}$	$V_3 = \frac{1}{m^2} [E(R) V(\frac{d-1}{y-1}) + (V(R) + V(A)) P^2 + E(A) V(\frac{c-1}{y-1})]$ $V_3^* = \frac{1}{m^2} [E(R) \frac{\tilde{p}_3(1-\tilde{p}_3)}{n-2} \frac{N-n+1}{N} + (V(R) + V(A)) \tilde{p}_3{}^2 + E(A) \frac{\tilde{p}_3(1-\tilde{p}_3)}{n-2} \frac{N-n+1}{N}]$ E stands for expected value

ON THE MULTIVARIATE POLYA DISTRIBUTION: A MODEL OF CONTAGION FOR DATA WITH MULTIPLE COUNTS

K. G. JANARDAN
G. P. PATIL
The Pennsylvania State University

SUMMARY

In this paper, the authors discuss certain chance mechanisms generating multivariate Polya (MPD) distribution, present different available forms of the probability function of the MPD, establish some interesting structural properties and a sampling property, discuss location of mode, certain inter-relations of MPD with other distributions and moment estimation of the parameters of MPD.

1. INTRODUCTION

The Polya urn scheme for a single variate case is well known as a model for contagion. The multivariate version does not seem to have received much attention. Steyn [6, 7] refers to it briefly while discussing what he calls as the multivariate hypergeometric series distributions. Mosimann [3] finds it of some use in pollen analysis under an equivalent form, seemingly not identified with the multivariate Polya distribution (MPD), however. In this paper, we propose to discuss certain structural properties and estimation problems for the MPD.

2. NOTATION AND TERMINOLOGY

To begin with, we shall discuss the notation, the definitions, and the terminology used in this paper. The s-dimensional vector $(x_1,x_2,\ldots,x_s)$ is denoted by $\underset{\sim}{x}$. The sum of the components of $\underset{\sim}{x}$ will be denoted by x. The multinomial coefficient will be denoted by

$$\frac{n!}{\prod_{i=o}^{s} x_i!} = \frac{n!}{(n-x)!\prod_{i=1}^{s} x_i!} = \binom{n}{\underset{\sim}{x}} . \tag{2.1}$$

The following notation (2.2) is based upon that of Bosch [1]

$$a^{(x,c)} = a(a + c)(a + 2c)\ldots(a + (x - 1)\ c) \tag{2.2}$$

$$c = -\ 1{:}a^{(x)} = a(a - 1)(a - 2)\ldots(a - x + 1) \tag{2.3}$$

$$c = +\ 1{:}a^{[x]} = a(a + 1)(a + 2c)\ldots(a + x - 1) \tag{2.4}$$

Clearly, the multinomial coefficient (2.1) can be expressed as

$$\binom{n}{\underset{\sim}{x}} = \binom{n - k}{\underset{\sim}{x} - \underset{\sim}{k}} \frac{n^{(k)}}{\prod_{i=1}^{s} x_i^{(k_i)}}, \tag{2.5}$$

and the factorial (2.2) as

$$a^{(x,c)} = a^{(k,c)}(a + kc)^{(x - k,c)}. \tag{2.6}$$

We recall

$$\Gamma(a) = \int_0^\infty e^{-t}t^{a - 1}\,dt \text{ for } a > 0; \tag{2.7}$$

$$\Gamma(a + 1) = a\Gamma(a); \tag{2.8}$$

$$a^{[x]} = \frac{\Gamma(a + x)}{\Gamma(a)}, \tag{2.9}$$

where x is a positive integer.

$$\frac{a^{[x]}}{x!} = (-1)^x \binom{-a}{x} = \binom{a + x - 1}{x}. \tag{2.10}$$

A discrete random variable x having the probability function (pf),

$$P(x) = \binom{n}{x} \frac{a^{(n - x,c)} b^{(x,c)}}{(a + b)^{(n, c)}}, \tag{2.11}$$

is said to have a univariate Polya distribution.

3. CHANCE MECHANISMS GENERATING MULTIVARIATE POLYA

The MPD arises in different ways. In this Section, we present three approaches that we have known so far.

3.1 Eggenberger-Polya scheme:

Steyn [6] derives the MPD by the following multivariate version of the Eggenberger-Polya scheme:

Consider a finite population consisting of N individuals of which $N_1, N_2, \ldots, N_s$ have respectively certain independent characteristics $A_1, A_2, \ldots, A_s$ and $N_0 = N \; \sum_{i=1}^{s} N_i$ have the characteristic "not $(A_1, A_2, \ldots, A_s)$". We draw individuals one-by-one with replacement, note the characteristic of the individual drawn, and add c individuals of the same characteristic before the next drawing is performed. We repeat this n times. If x_i denotes the number of individuals of character A_i in the sample drawn, then the random vector $\underset{\sim}{x} = (x_1, x_2, \ldots, x_s)$ has s-variate Polya distribution. Writing $\underset{\sim}{N} = (N_1, N_2, \ldots, N_s)$ the pf is given by

$$mp(\underset{\sim}{x}, \underset{\sim}{N}; c; n) = \binom{n}{\underset{\sim}{x}} \frac{\prod_{i=o}^{s} N_i(N_i + c)\ldots(N_i + (x_i - 1)c)}{N(N + c)\ldots(N + (n - 1)c)}, \tag{3.1}$$

where $x_i = 0,1,2,\ldots,n$ for $i = 0,1,2,\ldots,s$, so that $\sum_{i=o}^{s} x_i = n$, $n = 1,2,\ldots;$ $N_i = 1,2,\ldots,N$ for $i = 0,1,2,\ldots,s$, so that $\sum_{i=o}^{s} N_i = N$.

The MPD (for $c \neq 0$) is a "true contagious" distribution in the sense of Polya where each "favorable" event increases (or decreases) the probability of a future event.

For the MPD we shall use the notation $\underset{\sim}{x} \sim MP(\underset{\sim}{N};c;n)$ and alternatively we shall use

$$(x_1,x_2,\ldots,x_s) \sim MP(N, N_1, N_2,\ldots,N_s;c;n).$$

A probability distribution having the pf (3.1) can also be termed as Singular MPD considering it as a distribution (s + 1) dimensions, and for the SMPD we shall use the notation $(x_0,x_1,\ldots,x_s) \sim SMP(N_0,N_1,N_2,\ldots,N_s;c;n)$.

In what follows we deduce different forms of the pf of the MPD utilizing the notations and definitions of Section 2.

$$mp(\underset{\sim}{x},\underset{\sim}{r},n) = \binom{n}{\underset{\sim}{x}} \frac{\prod_{i=o}^{s} r_i(r_i + 1)\ldots(r_i + (x_i - 1)}{r(r + 1)\ldots(r + n - 1)}, \tag{3.2}$$

where $\frac{N_i}{c} = r_i$ for $c \neq 0$ and $r = \sum_{i=o}^{s} r_i = \frac{N}{c}$.

$$mp(\underset{\sim}{x},\underset{\sim}{N};c;n) = \binom{n}{\underset{\sim}{x}} \frac{N_o^{(n - x_1,c)} \prod_{i=o}^{s} N_i^{(x_i,c)}}{N^{(n,c)}} \tag{3.3}$$

$$mp(\underset{\sim}{x},\underset{\sim}{r},n) = \binom{n}{\underset{\sim}{x}} \frac{\prod_{i=o}^{s} r_i^{[x_i]}}{r^{[n]}} \tag{3.4}$$

$$mp(\underset{\sim}{x},\underset{\sim}{p}\,;\nu;n) = \binom{n}{\underset{\sim}{x}} \frac{\prod_{i=o}^{s} p_i(p_i + \nu)\ldots(p_i + (x_i - 1)\,\nu)}{1(1 + \nu)\ldots(1 + (n - 1)\,\nu)}, \tag{3.5}$$

where $p_i = \frac{N_i}{N} = \frac{r_i}{r}$ and $\nu = \frac{c}{N} = \frac{1}{r}$.

$$mp(\underset{\sim}{x},\underset{\sim}{p};\nu;n) = \binom{n}{\underset{\sim}{x}} \frac{\prod_{i=o}^{s} p_i^{(x_i,\nu)}}{1^{(n,\nu)}}, \tag{3.6}$$

where p_i can be interpreted as the probability of drawing an individual of character A_i in the first draw, and the parameter ν, as the measure of the degree of the "contagion."

$$mp(\underset{\sim}{x},\underset{\sim}{r},n) = \binom{n}{\underset{\sim}{x}} \frac{\Gamma(r)}{\Gamma(r+n)} \prod_{i=o}^{s} \frac{\Gamma(r_i + x_i)}{\Gamma(r_i)}, \tag{3.7}$$

where r_i and r are as defined earlier.

$$mp(\underset{\sim}{x},\underset{\sim}{r},n) = \frac{\prod_{i=o}^{s} \binom{-r_i}{x_i}}{\binom{-r}{n}} \tag{3.8}$$

Thus, in the above we have expressed the MPD in eight different forms. In connection with the univariate Polya distribution, Bosch [1] has remarked: "... different authors who have derived this distribution by different arguments have sometimes arrived at different expressions without recognizing that they were dealing with one and the same distribution." This remark can be applicable for the MPD also. Hence, we have tried to present different available forms here.

3.2 <u>Multinomial mixture with Dirichlet</u>:

Mosimann [3] obtains the MPD in the form (3.7) as a multinomial mixture with Dirichlet distribution and calls the distribution compound multinomial distribution, seemingly unaware of the identity in the probability functions.

Consider a production process by which we produce batches of size n. Suppose each unit produced has a probability p_i of being a defective one with the ith kind of defect. (Assume that each unit can have only one distinct defect out of s + 1 defects.) If x_i denotes the number with defects of ith kind contained in a batch of size n, then $\underset{\sim}{x} = (x_1,x_2,\ldots,x_s)$ will have the multinomial distribution:

$$m(\underset{\sim}{x} \mid \underset{\sim}{p},\ n) = \binom{n}{\underset{\sim}{x}} \prod_{i=o}^{s} p_i^{x_i}, \tag{3.9}$$

where $x_i = 0,1,2,\ldots,n$ for $i = 0,1,2,\ldots,s$; and $p_o = 1 - \sum_{i=1}^{s} p_i$.

Further, suppose we assume p's to be constant within a batch, but to vary randomly from batch to batch (and n being constant over all the batches) in keeping with a Dirichlet distribution

$$d(\underset{\sim}{p},\underset{\sim}{r},r_o) = \frac{\Gamma(r)}{\prod_{i=o}^{s} \Gamma(r_i)} \prod_{i=o}^{s} p_i^{r_i-1}, \tag{3.10}$$

where $r_i > 0$ for $i = 0,1,2,\ldots,s$ and $r = \sum_{i=o}^{s} r_i$.

Then the probability that a particular batch will contain x_1 of 1st kind defects, x^2 of 2nd kind,...,x_s of sth kind is given by

$$P(\underset{\sim}{x}) = \binom{n}{\underset{\sim}{x}} \frac{\Gamma(s)}{\prod_{i=o}^{s} \Gamma(r_i)} \int_{\mathcal{S}} \prod_{i=o}^{s} p_i^{x_i + r_i - 1} \, dp_1 \ldots dp_s,$$

where $\mathcal{S}$ denotes the s-dimensional simplex:

$$\{(p_1, p_2, \ldots, p_s) \mid 0 \le p_i \le 1, \sum_{i=1}^{s} p_i \le 1\}$$

Clearly, $P(\underset{\sim}{x}) = \binom{n}{\underset{\sim}{x}} \frac{\Gamma(r)}{\Gamma(r+n)} \prod_{i=o}^{s} \frac{\Gamma(r_i + x_i)}{\Gamma(r_i)}$. (3.11)

Mosimann [3] discusses a very interesting application of the MPD to a problem in pollen analysis.

3.3 <u>Conditional distribution of independent negative binomial variates given their sum:</u>

Let $x_0, x_1, \ldots, x_s$ be s + 1 independent random variables with x_i having the negative binomial distributions with parameters r_i and λ, $0 < \lambda < 1$, so that $\sum_{i=o}^{s} x_i$ also has the negative binomial distribution with parameters $r = \sum_{i=o}^{s} r_i$ and λ. Then the conditional distribution of $(x_1, x_2, \ldots, x_s) \mid \sum_{i=o}^{s} x_i = n$ is an MPD having the form (3.8). For, by definition,

$$P(x_1, x_2, \ldots, x_s \mid \sum_{i=o}^{s} x_i = n) = \frac{P(x_1, x_2, \ldots, x_s)}{P(\Sigma x_i = n)}$$

$$= \frac{\prod_{i=o}^{s} \binom{-r_i}{x_i} \lambda^{r_i} (1 - \lambda)^{x_i}}{\binom{-r}{n} \lambda^{r} (1 - \lambda)^{n}}$$

$$= \frac{\prod_{i=o}^{s} \binom{-r_i}{x_i}}{\binom{-r}{n}} .$$

We note that Σx_i is a sufficient statistic for λ and that the MPD arising as a conditional distribution given this sufficient statistic could be useful in developing a conditional test for λ.

4. DISTRIBUTIONAL PROPERTIES

It is interesting to observe that the MPD enjoys properties similar to the properties of the multinomial distribution.

If $(x_1,x_2,\ldots,x_s) \sim MP(N,N_1,N_2,\ldots,N_s;c;n)$,

then

(i) $(x_{i_1},x_{i_2},\ldots,x_{i_s}) \sim MP(N,N_{i_1},N_{i_2},\ldots,N_{i_s};c;n)$,

where $(i_1,i_2,\ldots,i_s)$ is a permutation of $(1,2,\ldots,s)$;

(ii) $(x_1,x_2,\ldots,x_k) \sim MP(N,N_1,N_2,\ldots,N_k;c;n)$,

where $1 \leq k \leq s$;

(iii) $(x_1 + x_2 +\ldots+ x_s) \sim MP(N, \sum_{i=1}^{s} N_i;c;n) \equiv P(N, \sum_{1}^{s} N_i;c;n)$,

where P stands for the univariate Polya distribution;

(iv) $(x_1,x_2,\ldots,x_s) \mid \sum_{1}^{s} x_i = m \sim SMP(N_1,N_2,\ldots,N_s;c;m)$, where SMP stands for the singular MPD;

(v) $(x_1,x_2,\ldots,x_k) \mid (x_{k+1},\ldots,x_s) \sim MP(N^*,N_1,\ldots,N_k;c;n^*)$,

where $N^* = N - \sum_{k+1}^{s} N_i$ and $n^* = n - \sum_{k+1}^{s} x_i$;

(vi) $(x_s \mid x_{s-1},x_{s-2},\ldots,x_1) \sim P(N^+,N_s;c;n^+)$,

where $N^+ = N - \sum_{1}^{s-1} N_i$ and $n^+ = n - \sum_{1}^{s-1} x_i$;

(vii) $z_1,z_2,\ldots,z_k) \sim MP(N,M_1,M_2,\ldots,M_k;c;n)$,

where $z_j = \Sigma_j x_i$, $M_j = \Sigma_j N_i$ with Σ_j standing for summation taken over $i \in G_j$ for $j = 1,2,\ldots,k$, such that $\bigcup_{j=1}^{k} G_j = \{1,2,\ldots,s\}$ $G_j \cap G_{j'} = \phi$ for $j \neq j' = 1,2,\ldots,k$; and

(viii) $(x_1,x_2,\ldots,x_s) \mid (z_1,z_2,\ldots,z_k) \sim k$ independent singular MPD's of the type, SMP $(\{M_{ij}\},c;z_j)$ for $j = 1,2,\ldots,k$, where $M_{ij} = N_i$ for $i \in G_j$ $j = 1,2,\ldots,k$.

In what follows we offer proofs of the above properties using the basic identity (4.1) available from the pf of the MPD. Steyn [7] has given the proofs of the properties (ii) and (v) using the probability-generating function of the MPD, which he has written in terms of a multivariate hypergeometric series.

$$\sum_{x_i=o}^{n} \frac{N_o^{(n-x,c)} \prod_{i=1}^{s} N_i^{(x_i,c)}}{(n-x)! \prod_{i=1}^{s} x_i!} = \frac{N^{(n,c)}}{n!} \tag{4.1}$$

$i = 1,2,\ldots,s$

$x = \Sigma x_i \leq n$

Proof of (i): Trivial.

Proof of (ii):

$$P(x_1,x_2,\ldots,x_k) = \frac{n! \prod_{i=1}^{k} N_i^{(x_i,c)}}{\prod_{i=1}^{k} x_i!\; N^{(n,c)}} \sum_{\substack{x_{k+1},\ldots,x_s \\ \sum_{k+1}^{s} x_i \leq \sum_{1}^{k} x_i}} \frac{N_o^{(n-x,c)} \prod_{k+1}^{s} N_i^{(x_i,c)}}{(n-x)! \prod_{k+1}^{s} x_i!}$$

$$= \frac{n! \prod_{i=1}^{s} N_i^{(x_i,c)}}{\prod_{i=1}^{k} x_i!\; N^{(n,c)}} \frac{\left(N - \sum_{1}^{k} N_i\right)^{(n-\sum_{1}^{k} x_i,c)}}{\left(n - \sum_{1}^{k} x_i\right)!}$$

Rearranging,

$$P(x_1,x_2,\ldots,x_k) = \frac{n!}{(n-y) \prod_{i=1}^{k} x_i!} \frac{M_o^{(n-y,c)} \prod_{i=1}^{k} N_i^{(x_i,c)}}{N^{(n,c)}},$$

where $M_o = N - \sum_{i=1}^{k} N_i; y = \sum_{i=1}^{k} x_i.$

Proof of (iii):

$$P(x_1+x_2+\ldots+x_s) = \sum_{\substack{x_1,x_2,\ldots,x_s \\ x_1+x_2+\ldots+x_s = y}} \frac{n!}{(n-y)! \prod_{i=1}^{s} x_i!} \frac{N_o^{(n-y,c)} \prod_{i=1}^{s} N_i^{(x_i,c)}}{N^{(n,c)}}$$

$$= \frac{n!\, N_o^{(n-y,c)}}{(n-y)!\, N^{(n,c)}} \sum_{\substack{x_1,\ldots,x_s \\ x_1+x_2+\ldots+x_s = y}} \frac{\prod_{i=1}^{s} N_i^{(x_i,c)}}{\prod_{i=1}^{s} x_i!}$$

$$= \binom{n}{y} \frac{N_o^{(n-y,c)} (\sum_1^s N_i)^{(n-y,c)}}{N^{(n,c)}}$$

<u>Proofs of (iv), (v) and (vi)</u>: Straightforward.

<u>Proofs of (vii)</u>:

First note $(x_{m+1}, x_{m+2}, \ldots, x_s) \sim MP(N, N_{m+1}, \ldots, N_s; c; n) \iff (x_{m+1}, x_{m+2}, \ldots, x_s, x_{s+1}) \sim SMP(N_{m+1}, N_{m+2}, \ldots, N_{s+1}; c; n)$, suppressing c and n for convenience, we can write,

$$(x_{m+1}, x_{m+2}, \ldots, x_s) \sim MP(N_{m+1}, N_{m+2}, \ldots, N_s; N_{s+1})$$

$$\iff (x_{m+1}, x_{m+2}, \ldots, x_{s+1}) \sim SMP(N_{m+1}, \ldots, N_{s+1}),$$

where $N_{s+1} = N - N_{m+1} - \ldots - N_s$ and $x_{s+1} = n - x_{m+1} - \ldots - x_s$.

Now, let $x_{s+2} = n - x_{m+1} - x_{m+2} - \ldots - x_s$

$$= x_1 + x_2 + \ldots + x_m$$

$$= z_1;$$

and $N_{s+2} = N - N_{m+1} - \ldots - N_s$

$$= N_1 + N_2 + \ldots + N_m$$

$$= M_1;$$

then $(x_{m+1}, x_{m+2}, \ldots, x_{s+2}) \sim SMP(N_{m+1}, N_{m+2}, \ldots, N_{s+1}, N_{s+2})$.

Applying property (i), we have

$$(x_{s+2}, x_{m+1}, \ldots, x_{s+1}) \sim SMP(N_{s+2}, N_{m+1}, \ldots, N_{s+1})$$

$$\iff (z_1, x_{m+1}, \ldots, x_s) \sim MP(M_1, N_{m+1}, \ldots, N_s; N_{s+1})$$

We can now repeat this process to bring in z_2 and so on, finally obtaining

$$(z_1, z_2, \ldots, z_k) \sim MP(M_1, M_2, \ldots, M_k; M_{k+1})$$

i.e. $(z_1, z_2, \ldots, z_k) \sim MP(N, M_1, M_2, \ldots, M_k; c; n)$.

<u>Proof of (viii)</u>: Straightforward.

5. MOMENT PROPERTIES

Steyn [6] obtains the first and second order moments of the MPD using the probability-generating function expressed in terms of the multivariate hypergeometric series. Mosimann [3] obtains the generalized factorial moment $\mu'[k_1,k_2,\ldots,k_s]$ of the MPD by taking the expectation of the generalized factorial moment $n^{(k)} p_1^{k_1} p_2^{k_2} \ldots p_s^{k_s}$ of the multinomial distribution. Bosch [1] obtains the factorial moments of the univariate Polya distribution using the structural property of the binomial coefficient $\binom{n}{x}$ and the factorial $a^{(x,c)}$. We obtain below the generalized factorial moment of the MPD, using the structural property of the multinomial coefficient $\binom{n}{x}$ and the factorial $a^{(x,c)}$. We obtain below the generalized factorial moment of the MPD, using the structural property of the multinomial coefficient $\binom{n}{x}$ and that of the factorial $a^{(x,c)}$ given in (2.5) and (2.6), respectively.

Using (2.5) and (2.6), we write (3.3) as

$$mp(\underset{\sim}{n},\underset{\sim}{N};c;n) = \frac{n^{(k)}}{\prod_{i=1}^{s} x_i^{(k_i)}} \binom{n-k}{\underset{\sim}{x}-\underset{\sim}{k}} \frac{N_o^{(n-x,c)} \prod_{i=1}^{s} N_i^{(k_i,c)} \prod_{i=1}^{s} (N_i+k_i,c)^{(x_i-k_i,c)}}{N^{(k,c)} (N+k,c)^{(n-k,\,c)}}$$

$$= \frac{n^{(k)}}{\prod_{i=1}^{s} x_i^{(k_i)}} \frac{\prod_{i=1}^{s} N_i^{(k_i,c)}}{N^{(k,c)}} \, mp(\underset{\sim}{x}-\underset{\sim}{k},\ \underset{\sim}{N}+\underset{\sim}{k}\ c;\ c;\ n-k). \quad (5.1)$$

Hence,

$$\mu'[k_1,k_2,\ldots,k_s] = E[\prod_{i=1}^{s} x_i^{(k_i)}]$$

$$= \sum_{x_i=o}^{n} \prod_{i=1}^{s} x_i^{(k_i)} \, mp(\underset{\sim}{x},\underset{\sim}{N};c;n)$$

$$i = 1,2,\ldots,s$$

$$= \frac{n^{(k)} \prod_{i=1}^{s} N_i^{(k_i,c)}}{N^{(k,c)}} \sum_{x_i=k_i}^{n} mp(\underset{\sim}{x}-\underset{\sim}{k},\underset{\sim}{N}+\underset{\sim}{k}c;c;n-k) \quad (5.2)$$

$$= \frac{n^{(k)} \prod_{i=1}^{s} N_i^{(k_i,c)}}{N^{(k,c)}}, \tag{5.3}$$

since the multiple sum in (5.2) is unity as it represents the sum of the pf of an MPD with corresponding parameters.
(5.3) can be reduced to

$$\mu'[k_1,k_2,\ldots,k_s] = \frac{n^{(k)} \prod_{i=1}^{s} (\frac{N_i}{c} + k_i - 1)^{(k_i)}}{(\frac{N}{c} + k - 1)^{(k)}} \tag{5.4}$$

and also to

$$\mu'[k_1,k_2,\ldots,k_s] = \frac{n^{(k)} \prod_{i=1}^{s} p_i^{(k_i,\nu)}}{1^{(k,\nu)}}. \tag{5.5}$$

When s = 1:

$$\mu'[k_1] = \frac{n^{(k_1)} p_i^{(k_1,\nu)}}{1^{(k_1,\nu)}} \tag{5.6}$$

The moments and cumulants now follow from the known relations between factorial moments, moments and cumulants. The expected value, the variance of x_i, $cov(x_i,x_j)$, and the correlation coefficient of x_i and x_j are given below in terms of p_i, q_i ν and n:

$\mu_i = E(x_i) = np_i.$

$\sigma_{ii} = V(x_i) - np_i q_i \frac{1 + n\nu}{1 + \nu}$, where $q_i = 1 - p_i$.

$\sigma_{ij} = Cov(x_i,x_j) = - np_i q_i \frac{1 + n\nu}{1 + \nu}.$

$\rho_{ij} = Corr(x_i,x_j) = - \sqrt{\frac{p_i pj}{q_i q_j}}.$

The variance-covariance matrix is

$$\Sigma_{mp} = \begin{bmatrix} \sigma_{11} & \cdots & \sigma_{1s} \\ \sigma_{s1} & \cdots & \sigma_{ss} \end{bmatrix} = \frac{1 + n\,\nu}{1 + \nu} \Sigma_m = R\,\Sigma_m, \tag{5.7}$$

where Σ_m denotes the variance-covariance matrix of a multinomial and $R = \frac{1 + n\nu}{1 + \nu}$.

The generalized variance of the MPD is $|\Sigma_{mp}| = R^s |\Sigma_m|$. From property (vi) of Section 4, it follows that the regression of x_s on the other variables $x_1, x_2, \ldots, x_{s-1}$ is linear, and is given by

$$E(x_s | x_{s-1}, x_{s-2}, \ldots, x_1) = np_s - \frac{p_s}{p_o + p_s}\{(x_1 - np_1) + (x_2 - np_2) + \ldots + (x_{s-1} - np_{s-1})\},$$

which shows that the regression function depends only on the sum of the conditioning variables and not on their individual values. Next, we have the conditional variance given by

$$V(x_s | x_{s-1}, x_{s-2}, \ldots, x_1) = (n - x_1 - x_2 - \ldots - x_{s-1}) \frac{p_o p_s}{(p_o + p_s)^2} \frac{1 + (n - x_1 - x_2 - \ldots - x_{s-1})\nu}{1 + \nu},$$

which shows that the regression is heteroscedastic.

6. A SAMPLING PROPERTY

Consider the Eggenberger-Polya sampling scheme discussed in Section 3.1. First draw a sample of n_1 individuals and then continue to draw a second sample of n_2 individuals. If $\underset{\sim}{x}_1 = (x_{11}, x_{12}, \ldots, s_{1s})$ and $\underset{\sim}{x}_2 = (x_{21}, x_{22}, \ldots, x_{2s})$ are the random vectors denoting the numbers of individuals of the various characteristics observed in these two samples, then

$$x_1 \sim MP(\underset{\sim}{N}, c; n_1)$$

and $\underset{\sim}{x}_2 | \underset{\sim}{x}_1 \sim MP(\underset{\sim}{N} + \underset{\sim}{x}_1 c, c; n_2)$.

Suppose if $n = n_1 + n_2$ and $\underset{\sim}{x} = \underset{\sim}{x}_1 + \underset{\sim}{x}_2$, then $\underset{\sim}{x} \sim MP(\underset{\sim}{N}, c; n)$. For the joint distribution $\underset{\sim}{x}_1$ and $\underset{\sim}{x}_2$ is

$$P(\underset{\sim}{x}_1, \underset{\sim}{x}_2) = mp(\underset{\sim}{x}_1, \underset{\sim}{N}; c; n_1)\, mp(\underset{\sim}{x}_2, \underset{\sim}{N} + \underset{\sim}{x}_1 c; c; n_2)$$

$$= \binom{n_1}{\underset{\sim}{x}_1} \frac{N_o^{(n_1 - x_1, c)} \prod_{i=1}^{s} N_i^{(x_{1i}, c)}}{N^{(n_1, c)}}.$$

$$\binom{n_2}{\underset{\sim}{x}_2} \frac{(N_o + (n_1-x_1)c)^{(n_2-x_2,c)} \prod_{i=1}^{s} (N_i+x_{1i}c)^{(x_{2i},c)}}{(N + n_1c)^{(n_2,c)}},$$

where $x_1 = \sum_{i=1}^{s} x_{1i}$ and $x_2 = \sum_{j=1}^{s} x_{2j}$.

Hence,

$$P(\underset{\sim}{x}) = P(\underset{\sim}{x}_1 + \underset{\sim}{x}_2)$$

$$= \frac{N_o^{(n_1 - x_1,\ c)} (N_o + (n_1 - x_1)c)^{(n_2 - x_2,\ c)}}{N^{(n_1,c)} (N + n_1c)^{(n_2,c)}} \cdot$$

$$\sum_{\substack{\underset{\sim}{x}_1+\underset{\sim}{x}_2 = \underset{\sim}{x} \\ n_1+n_2 = n}} \binom{n_1}{\underset{\sim}{x}_1}\binom{n_2}{\underset{\sim}{x}_2} \prod_{i=1}^{s} N_i^{(x_{1i},c)} (N_i + x_{1i}c)^{(x_{2i},c)}$$

$$= \frac{N_o^{(n - x,c)} \prod_{i=1}^{s} N_i^{(x_i,c)}}{N^{(n,c)}} \sum_{\substack{\underset{\sim}{x}_1+\underset{\sim}{x}_2 = \underset{\sim}{x} \\ n_1+n_2 = n}} \binom{n_1}{\underset{\sim}{x}_1} \binom{n_2}{\underset{\sim}{x}_2},$$

where $x_{1i} + x_{2i} = x_i$

$$= \binom{n}{\underset{\sim}{x}} \frac{N_o^{(n - x)} \prod_{i=1}^{s} N_i^{(x_i,c)}}{N^{(n,c)}} .$$

7. INTERRELATIONS WITH OTHER DISTRIBUTIONS

In this Section we shall show that several multivariate distributions, both discrete and continuous, are related to the MPD either as particular cases or as limiting distributions.

a) Bose-Einstein uniform distribution.

This distribution is obtained by taking $p_i = \frac{1}{1+s}$ for $i = 0,1,2,\ldots,s$ and $\nu = \frac{1}{1+s}$ in (3.5) and is given by the pf

$$mp(\underset{\sim}{x};n) = \frac{1}{\binom{n+s}{s}},\quad x_i = 0,1,2,\ldots,n \text{ for } i = 0,1,\ldots,s \text{ so}$$

that $\sum_{i=o}^{s} x_i = n.$

This distribution has been discussed in Särndal [5] as a prior distribution in the derivation of certain distributions through Bayes approach.

b) Multinomial distribution.

We get this distribution from the MPD by taking $\nu = o$. Since $p_i^{(x_i,o)} = p_i^{x_i}$, we have from (3.6),

$$mp(\underset{\sim}{x};\underset{\sim}{p};n) = \binom{n}{\underset{\sim}{x}} \prod_{i=o}^{s} p_i^{x_i},$$

where $x_i = 0,1,2,\ldots,n$ for $i = 0,1,2,\ldots,s$ so that $\sum_{i=o}^{s} x_i = n.$

This is also evident from Eggenberger-Polya's sampling scheme when $c = o$. In that case the individual drawn is simply replaced, but no more additional individuals are added or removed.

c) Multivariate hypergeometric distribution.

We get this distribution by taking $c = -1$ in (3.3), or by taking $\nu = -\frac{1}{N}$ and $p_i = \frac{N_i}{N}$ in (3.5), we then have the pf

$$mp(\underset{\sim}{x};\underset{\sim}{N};n) = \frac{\prod_{i=o}^{s} \binom{N_i}{x_i}}{\binom{N}{n}},$$

where $x_i = 0,1,2,\ldots,n$ for $i = 0,1,2,\ldots,s$ so that $\sum_{i=o}^{s} x_i = n.$

d) Multivariate negative hypergeometric distribution.

We get this by taking $c = +1$ in (3.3), or $\nu = -\frac{1}{N}$ and $p_i = \frac{N_i}{N}$ in (3.5), we then have the pf

$$mp(\underset{\sim}{x};\underset{\sim}{N};n) = \frac{\prod_{i=o}^{s} \binom{-N_i}{x_i}}{\binom{-N}{n}}.$$

e) Multiple Poisson distribution.

The multiple Poisson distribution is obtained, as usual, as a limit of the multinomial distribution. We first take $\nu = o$ in (3.5) and then take limit as $n \to \infty$, $p_i \to o$ such that $np_i \to \lambda_i$. We have then the pf

$$mp(\underset{\sim}{x};\underset{\sim}{p};n) \to \prod_{i=o}^{s} e^{-\lambda_i} \frac{\lambda_i^{x_i}}{x_i!}$$

f) Independent negative binomial distributions.

The MPD tends to independent negative binomial distributions with parameters (r_i, λ), $i = 1,2,\ldots,s$, respectively, if $r \to \infty$, $n \to \infty$ with $r_1, r_2, \ldots, r_s$ remaining fixed such that $\frac{r}{r+n} \to \lambda$, $o < \lambda < 1$. We have from (3.7)

$$mp(\underset{\sim}{x};\ \underset{\sim}{r};\ n) = A \prod_{i=1}^{s} \frac{\Gamma(r_i + x_i)}{\Gamma(r_i)\ x_i!} ,$$

where $A = \frac{\Gamma(r_o + x_o)}{\Gamma\ (r_o)x_o!} \cdot \frac{n!\ \Gamma(r)}{\Gamma(r + n)} = \frac{(r_o + x_o - 1)!\ n!\ (r - 1)!}{(r_o - 1)!\ x_o!\ (r + n - 1)!}$.

Applying Stirling's approximations for the factorials and taking limit of A as $r \to \infty$, $n \to \infty$, such that $\frac{r}{r+n} \to \lambda$, where $r_1, r_2, \ldots, r_s$ are kept constant, we can easily see that

$$A \to \lambda^{\sum_1^s r_i} (1 - \lambda)^{\sum_1^s x_i} .$$

Thus $mp(\underset{\sim}{x};\ \underset{\sim}{r};\ n) \to \prod_{i=1}^{s} \frac{\Gamma(r_i + x_i)}{\Gamma(r_i)\ x_i!} \lambda^{r_i} (1 - \lambda)^{x_i}$.

g) Dirichlet distribution.

The MPD tends to the Dirichlet distribution as n tends to infinity such that $\frac{x_i}{n}$ $(i = 1,2,\ldots,s)$ remains finite and $r, r_1, r_2, \ldots, r_s$ remain fixed.

Since $\frac{\Gamma(r_i + x_i)}{x_i!} = (r_i - 1)^{(x_i)}$, we have from (3.7)

$$mp(\underset{\sim}{x};\underset{\sim}{r};n) = \frac{\Gamma(r)}{\prod_{i=o}^{s} \Gamma(r_i)} \ \frac{\prod_{i=o}^{s} (r_i - 1)^{(x_i)}}{(r - 1)^{(n)}}$$

$$= \frac{\Gamma(r)}{\prod_{i=o}^{s} \Gamma(r_i)} \ \frac{\prod_{i=o}^{s} (\frac{x_i}{n} + \frac{r_i - 1}{n})\ldots(\frac{x_i}{n} + \frac{1}{n})}{n^s(1 + \frac{r-1}{n})\ldots(1 + \frac{1}{n})} .$$

Putting $\frac{x_i}{n} = p_i$, $dx_i = ndp_i$, we obtain

$$mp(\underset{\sim}{x};\underset{\sim}{r};n) \to \frac{\Gamma(r)}{\prod_{i=o}^{s} \Gamma(r_i)} \prod_{i=o}^{s} p_i^{r_i - 1} .$$

h) Multivariate normal distribution.

The MPD tends for large values of $|n|$, $|r_1|,\ldots,|r_s|$ and $|r_o + n - 1|$ (but of the same order as $O(n)$) to the multivariate normal distribution (Steyn [7]).

8. MODE OF THE MPD

Following Finucan [2], we introduce the following definitions:

1) $\underset{\sim}{y}$ is [a,b] neighbor of $\underset{\sim}{x}$ if $y_a = x_a + 1$, $y_b = x_b - 1$ and $y_i = x_i$ for $i \neq a,b$.

2) $\underset{\sim}{x}$ is a strict local mode if its probability exceeds that of all of its neighbors.

3) A joint mode is one which has an equiprobable neighbor.

Using the definitions (1) and (2) we obtain the conditions for a strict local mode for the SMPD as follows:

$$\frac{P(x_1,x_2,\ldots,x_i + 1,\ldots,x_j - 1,\ldots,x_s)}{P(x_1,x_2,\ldots,x_s} < 1 \Rightarrow \frac{N_i+x_ic}{x_i+1} < \frac{N_j+(x_j-1)c}{x_j},$$

which in turn reduces to

$$\frac{x_i + 1}{\frac{N_i}{c} - 1} > \frac{x_j}{\frac{N_j}{c} - 1}, \text{ for all } i,\ j, \tag{8.1}$$

where $c > o$.

Using definitions (1) and (3) we obtain the conditions for a point to be a joint mode for the SMPD as follows:

Let $\underset{\sim}{x}$ be a joint mode and $\sum_{i=1}^{s} x_i = n$.

Suppose one of its equally probable neighbors is [a,b] neighbor; then

$$\frac{x_a + 1}{\frac{N_a}{c} - 1} > \frac{x_b}{\frac{N_b}{c} - k} \Rightarrow \frac{x_b + 1}{\frac{N_b}{c} - 1} > \frac{x_a}{\frac{N_a}{c} - 1},$$

where $c > o$. Also no neighbor is more probable than $\underset{\sim}{x}$; hence,

$$\frac{x_i + 1}{\frac{N_i}{c} - 1} \geq \frac{x_j}{\frac{N_j}{c} - 1}, \text{ for all } i, j = 1,2,\ldots,s. \tag{8.2}$$

The conditions (8.1) and (8.2) correspond to the conditions given by Finucan [2] for a strict local mode and for a point to be a joint mode, respectively, for the multinomial distribution. Hence, mode(s) may be obtained by finding a constant K so that $\underset{\sim}{x}$ is defined by $x_i = [K(N_i - c)]$ such that $\sum_1^s x_i = n$, where [a] is the integral part of a.

If K is unique, then there are many acceptable points $\underset{\sim}{x}$ such that $\sum_{i=1}^{s} x_i = n$ which form a set of joint modes. If K is not unique, then we get a unique $\underset{\sim}{x}$ such that $\Sigma x_i = n$, which would be the unique mode of the MPD.

9. ESTIMATION OF THE PARAMETERS

We consider here the moment estimators of the parameters of the MPD under the assumptions c > o, n is a known positive integer and N is unknown. If N is known, the problem becomes trivial, and in this case the estimators are given by

$$\hat{N}_i = \frac{N}{n} \bar{x}_i \quad i = 1,2,\ldots,s. \tag{9.1}$$

(That the $\hat{N}_i$'s are the maximum likelihood estimators of N_i will be shown elsewhere.)

To obtain the estimators of the parameters $N_0, N_1, \ldots, N_s$ by the method of moments on the basis of a random sample of size m with x_{ij} $i = 1,2,\ldots,s$, $j = 1,2,\ldots,m$ taken on $(x_1, x_2, \ldots, x_s) \sim MP(N, N_1, N_2, \ldots, N_s; c; n)$, we have from Mosimann [3] the moment equations:

$$\bar{x}_i = n \frac{\hat{r}_i}{\Sigma r_i} \quad i = 1,2,\ldots,s$$

$$\hat{r}_0 + \hat{r}_1 + \ldots + \hat{r}_s = \frac{n - \hat{R}}{\hat{R} - 1},$$

which can be written as

$$\hat{N}_o + \hat{N}_1 + \ldots + \left(\frac{x_i - n}{\bar{x}_i}\right) \hat{N}_i + \ldots + \hat{N}_s = 0 \quad i = 1,2,\ldots,s \tag{9.2}$$

$$\hat{N}_0 + \hat{N}_1 + \ldots + \hat{N}_s = \frac{c(n - \hat{R})}{\hat{R} - 1}, \tag{9.3}$$

where $\hat{R}$ is an estimator of $R = \frac{1 + n\nu}{1 + \nu} = \frac{\sum_{i=o}^{s} N_i + nc}{\sum_{i=o}^{s} N_i + c}$.

(9.2) and (9.3) yield the moment estimators as

$$\hat{N}_i = \bar{x}_i \left[\frac{c(n - \hat{R})}{n(\hat{R} - 1)}\right] \tag{9.4}$$

$$\hat{N}_o = (n - \sum_{i=1}^{s} \bar{x}_i) \left[\frac{c(n - \hat{R})}{n(\hat{R} - 1)}\right] \tag{9.5}$$

There are many possible ways of getting an estimate of R. Mosimann [3] suggests the estimate of R to be based on either a ratio of a number of variances and covariances, or as

$$\hat{R} = \left(\frac{|W_{mp}|}{|W_m|}\right)^{\frac{1}{s}}, \tag{9.6}$$

where W_{mp} and W_m are consistent estimates of Σ_{mp} and Σ_m, or an estimate based on one variable only. We propose here a consistent estimate of R based on the variance of the sum of the components. From property (iii) of Section 4, we have $\Sigma x_i \sim P(N, \Sigma N_i; c; n)$, and hence its variance equals $RnP(1 - P)$, where

$$P = \frac{\Sigma N_i}{N}.$$

Thus,

$$\hat{R} = \frac{n S_x^2}{\bar{x}(n - \bar{x})}, \tag{9.7}$$

where S_x^2 is the sample variance and $\bar{x}$ the mean of the total number of observations.

REFERENCES

[1] Bosch, A. J. 1963. "The Polya distribution," Statistica Neerlandica 17. nr. 3, 201-13.

[2] Finucan, H. M. 1964. "The mode of a multinomial distribution." Biometrika 51:513-17.

[3] Mosimann, J. E. 1962. "On the compound multinomial distribution, the multivariate β-distribution and correlations." Biometrika 49:65-82.

[4] Patil, G. P., and Joshi, S. W. 1968. A Dictionary and Bibliography of Discrete Distributions. pp. 268. Publisher, Oliver and Boyd, Edinburgh.

[5] Särndal, C. E. 1964. "Derivation of a class of frequency distributions via Bayes Theorem." J. Amer. Statist. Assoc. 59:1042-53.

[6] Steyn, H. S. 1951. "On discrete multivariate probability functions." Proc. Koninklijke Nederlands Akademie Van Wetenschappen, Series A, 54, No. 1, 23-30.

[7] ———. 1955. "On discrete multivariate probability functions of hypergeometric type." Nederl. Akad. Wetensch. Proc. Series A, 58, 588-95.

DISCRETE DISTRIBUTIONS IN MARKET RESEARCH

C. CHATFIELD
Bath University of Technology
England

SUMMARY

The applications of discrete distributions in such fields as biology, psychology and ecology are well known. The purpose of this paper is to draw attention to the increasing use of discrete distributions in the study of consumer purchasing behavior. The models discussed in this paper are concerned with the purchase of nondurable consumer products such as detergents, breakfast cereals, margarine, petrol, etc.

1. INTRODUCTION

The applications of discrete distributions in such fields as biology, psychology and ecology are well known. The purpose of this paper is to draw attention to the increasing use of discrete distributions in the study of consumer purchasing behavior.

The models discussed in this paper are concerned with the purchase of nondurable consumer products such as detergents, breakfast cereals, margarine, petrol, etc. The data is obtained from a continuous consumer panel in which a random sample of consumers keep a record of their own purchases over an extended time period. The analyses are made over time periods which are multiples of one week—often 4, 12 or 13 weeks. Time periods shorter than one week cannot be analyzed by the methods discussed here since purchases of an individual consumer on successive days may be negatively correlated.

The study is restricted to one brand at a time, and, in the first part of the paper, to one size pack of a specific brand. Some remarks on this restriction are included in Section 6.

The first model described in this paper involves the negative binomial distribution (abbreviated NBD) and also the logarithmic series distribution (LSD), while second model involves the beta-binomial distribution (BBD).

There are two broad objectives in setting up these models: first, that they should give a good description of different aspects of purchasing behavior, and second, that they should enable predictions to be made. As regards the first objective, the marketing man had very little generalized knowledge about consumer purchasing behavior prior to this type of investigation. For example, if half the buyers of a particular brand account for 85% of all purchases, he would like to know if such behavior is in some way 'typical.' For the second objective, he would like to use data from a given time period to predict various aspects of the behavior of the sample in subsequent time periods. These predictions will depend on the assumption that the market is stationary. This means that there is no overall change in the level of purchasing from one period to the next, even though the purchases of individual consumers may vary. If the market is, in fact, nonstationary, as will be the case during a promotional campaign, for example, then the stationary predictions can be used as a yardstick to assess the observed changes in purchasing behavior.

2. THE NSD/LSD MODEL

One way of presenting the data obtained from a consumer panel is to construct the frequency distribution of purchases by finding the number of consumers in the sample who bought 0,1,2,..., or, in general, r units (or packets) of the given product in some specified time period. An example of such a distribution is given in Table 1.

Ehrenberg [5] noted that the negative binomial distribution gives a good fit to distributions of this type. Various models have been proposed to generate the NBD, but further study showed that a compound Poisson model was appropriate in this case, and so Ehrenberg proposed the following model for one pack size of a specific brand:

(a) The purchases of a particular consumer in successive (equal) time periods are independent and follow a Poisson distribution with a constant mean.

(b) This average long-run rate of buying will vary from consumer to consumer, the distribution being a Pearson type III (or gamma) distribution.

As mentioned before, the time periods referred to here must be multiples of one week. Also notice that the model assumes that the market is stationary.

The restriction to one pack size of a specific brand is necessitated by the complications which would arise from trying to mix purchases of large, medium and small sizes to give an integer valued distribution. However it is sometimes desirable to analyze all the purchases of a brand at the same time rather than just analyzing purchases of a specific pack size. This can be done in two ways, either by letting the NBD model apply to purchasing occasions rather than purchases, or by means of the model described in Section 4. But the remainder of this Section is concerned with purchases of a specific pack size where the product is marketed in more than one size.

An immediate consequence of the above compound Poisson model is that the frequency distribution of purchases will follow a negative binomial distribution. It is convenient to express the p.d.f. of the Pearson type III distribution in the following form:

$$f_P(\lambda) = \frac{e^{-\lambda k/m}\lambda^{k-1}}{\left(\frac{m}{k}\right)^k \Gamma(k)} \quad . \quad \begin{array}{l} 0 \le \lambda \le \infty \\ m > 0. \ k > 0. \end{array}$$

Then the proportion of the population who purchase r units of the brand in a particular time period is given by

$$P(r) = \int_0^\infty f_p\ (\lambda)\ \frac{e^{-\lambda}\ \lambda^r}{r!}\ d\lambda$$

$$= (1 + \frac{m}{k})^{-k}\ \frac{\Gamma(k+r)}{\Gamma(k).r!} \cdot \left(\frac{m}{m+k}\right)^r \qquad r = 0,1,2,\ldots,\infty$$

The two parameters of the NBD are the mean m and the exponent k. The cumulative generating function of the NBD is given by

$$k\ \ln[1 - \frac{m}{k}\ (e^t - 1)].$$

Thus the mean and variance are m and $m(1 + \frac{m}{k})$, respectively.

Methods of estimation are discussed by Anscombe [1]. The method customarily used in market research is to equate the observed mean and proportion of non-buyers to their theoretical values. This is called fitting by mean and zeros. If the sample mean is denoted by $\bar{x}$ and the proportion of buyers by b, we have

$$\bar{x} = \hat{m} \text{ and } 1 - b = (1 + {}^{\bar{x}}/\hat{k})^{-\hat{k}}.$$

This method has a number of advantages. First, the sample mean is the maximum likelihood estimate of m, and the estimate of k is nearly 100% efficient for the reverse J-shaped distributions which commonly occur here. Second, the maximum likelihood estimate of k (which is very tedious to derive) is impossible to obtain unless the original frequency distribution is known. Sometimes the values of $\bar{x}$ and b is the only information available. Third, it can be shown that this method of estimation ensures that repeat-buying predictions for the sample (which are discussed in the next Section) will be unbiased.

The zeros estimate of k, or more usually of a = m/k, can be obtained by iteration. Given the sample mean $\bar{x}$ and the proportion of buyers b, the estimate $\hat{a}$ is given implicitly by

$$1 - b = (1 + \hat{a})^{-\bar{x}/\hat{a}}.$$

The solution of this equation is discussed by Chatfield [4], who tabulates the value of a for different values of $c = -\bar{x}/\ln(1 - b)$. Note that if $c \leq 1$, then an inadmissible (i.e., negative) value of a will result. Values of a are plotted in Figure 1 for different values of b and $w = \bar{x}/b$. The quantity w is the average rate of buying per buyer and is an extremely useful statistic in market research. Its importance will be fully appreciated after reading the remainder of this Section.

A NBD was fitted to the data in Table 1 and the fit, as judged by eye or by a χ^2 goodness-of-fit test, is very good.

Table 1. Example of the fit of the NBD and LSD to an observed frequency distribution.
Brand K, 4 weeks.

Number of units	Observed Frequency	NBD	LSD
0	1671	1671.0	1671.0
1	43	45.2	45.5
2	19	16.3	16.1
3	9	7.7	7.6
4	2	4.1	4.0
5	3	2.3	2.3
6	1	1.3	1.3
7	0	0.8	0.8
8	0	0.5	0.5
9	2	0.3	0.3
10+	0	0.5	0.6
	$\bar{x} = 0.089$ $b = 0.045$ $w = 1.96$ $s^2 = 0.28$	$m = 0.089$ $k = 0.039$ $\sigma^2 = 0.29$	$q = 0.707$ $\sigma^2 = 0.29$

A large number of distributions from a variety of product fields was fitted by a NBD and, generally speaking, a good fit was obtained. A quick method of assessing the goodness-of-fit is to compare the observed and theoretical variances. Most of the distributions checked in this way gave good agreement. But for some heavily bought products the theoretical variance is systematically higher than the observed variance. This is the so-called variance discrepancy discussed by Chatfield et al. [2]. Fortunately, even when the variance discrepancy occurs, most other aspects of the NBD model seem to be valid, and so the many useful consequences of the model can be utilized.

Before discussing practical applications of the NBD model,

a modification will be introduced which enables many results to be simplified. If the distribution of purchases in some time period can be described by a NBD, and if the NBD parameter k is less than about 0.15, then the distribution of purchases, excluding the non-buyers, can be approximated by a distribution called the logarithmic series distribution.

The LSD can be written as:

$$P(r) = -q^r/r \ln(1 - q) \qquad r = 1,2,\ldots,\infty$$
$$0 < q < 1.$$

The moment-generating function is given by

$$\ln(1 - qe^t)/\ln(1 - q),$$

so that the mean and variance are given by

$$\mu = -q/(1 - q)\ln(1 - q)$$

$$\sigma^2 = -q[1 + \frac{q}{\ln(1 - q)}]/(1 - q)^2 \ln(1 - q).$$

The properties of the LSD are discussed, for example, by Patil et al. [9] and Williamson and Bretherton [12].

Note particularly that this distribution describes the proportion of <u>buyers</u> who purchase r units in the time period. In order to obtain the complete distribution, we must introduce a second parameter b which denotes the proportion of buyers.

Then $P(0) = 1 - b$

$$P(r) = -bq^r/\ln(1 - q) \times r. \qquad r = 1,2,\ldots,\infty.$$

The theoretical proportion of buyers is always estimated by the observed proportion of buyers. For simplicity both quantities will be denoted by b, since no confusion is likely to arise. The maximum likelihood estimate of q can be obtained by equating the observed and theoretical means of the distribution of buyers. As the observed mean is w, the estimate of q is given implicitly by

$$w = \hat{q}/(1 - \hat{q}) \ln(1 - \hat{q})$$

Williamson and Bretherton [12] tabulate values of q for different values of w. Alternatively, Chatfield [4] tabulates the value of $a_L = q/(1 - q)$ for different values of w, as there are some advantages to using a_L rather than q. Values of a_L can be obtained, in fact, from Figure 1 by putting b = 0.

A compound Poisson model for the LSD, along the lines of the NBD model, is discussed briefly in Chatfield et al. [2]. The LSD was originally derived by Fisher [8] who showed that if the

NBD parameters m and k tend to zero in such a way that their ratio $a = m/k$ tends to a nonzero limit, then the nonzero portion of the NBD tends to the LSD. The limit of a is the quantity a_L.

The LSD can also be obtained as a limiting case of the NBD by letting the proportion of non-buyers tend to 1, while keeping the mean of the truncated distribution constant.

$$(1 + a)^{-m/a} = 1 - b$$

$$-\frac{m}{a} \ln(1 + a) = \ln(1 - b)$$

$$= b - b^2/2 + b^3/3 \; \ldots .$$

$$-\frac{w}{a} \ln(1 + a) = 1 - b/2 + b^2/3 \; \ldots .$$

As $b \to 0$ with w constant, the value of a tends to a limit a_L which can be obtained from the equation

$$w = - a_L/\ln(1 + a_L).$$

The parameters $m(= wb)$ and $k(= m/a)$ will both tend to zero. The recurrence relationship for the NBD is given by

$$P(r) = \left(\frac{a}{1 + a}\right) \left(1 - \frac{a - m}{ar}\right) P(r - 1).$$

As $b \to 0$ with w constant, we have

$$P(r) \to \left(\frac{a_L}{1 + a}\right) \left(\frac{r - 1}{r}\right) P(r - 1),$$

which is the recurrence relationship for the LSD. The fact that a tends to a nonzero limit as $b \to 0$ with w constant is clearly visible in Figure 1.

An example of a distribution fitted by both a NBD and a LSD is given in Table 1. The two distributions are very similar. The question then arises as to what empirical distributions will give a fitted NBD and LSD which are quite alike. Given values for w and b, it is possible to estimate the NBD parameters m and a and also to estimate the LSD parameter q (or a_L). Then the characterictics of the two fitted distributions can be compared for different values of w and b. Since the number of non-buyers is the same for both distributions, we need only compare the truncated distributions. The variance of the LSD can be expressed simply as $w(1 + a_L - w)$ or $w\left(\frac{1}{1 - q} - w\right)$. The variance of the truncated NBD can be expressed as $w(1 + a + m - w)$.

Thus the ratio of the standard deviations is given by

$$\sqrt{\frac{1 + a_L - w}{1 + a + m - w}}\ .$$

Values of this ratio are given in Table 2.

Table 2. Values of $\sqrt{\frac{1 + a_L - w}{1 + a + m - w}}$ for different values of w and b.

		b 0.1	0.2	0.3	0.4
	1.6	1.033	1.072	1.116	1.171
	2.4	1.032	1.068	1.111	1.162
	3.2	1.031	1.066	1.107	1.156
w	4.0	1.031	1.065	1.105	1.153
	7.0	1.029	1.063	1.101	1.146
	10.0	1.029	1.062	1.100	1.145
	15.0	1.029	1.061	1.098	1.142

As was to be expected, the ratio depends mainly on the value of b. When b = 0.1 the difference is around 3%, but when b = 0.4 the difference is around 15%. Of course the difference between the standard deviations of the complete distributions will be much smaller, because the distributions are then dominated by the non-buyers.

3. SOME PREDICTIONS

Given information in some basic time period, such as 4 or 12 weeks, the NBD/LSD model enables us to predict how the sample will behave in subsequent time periods. Under stationary conditions, it is easy to show that the distribution of purchases in a time period T times as long will also be a NBD with parameters

$$m_T = Tm$$

and $$k_T = k$$

Thus the proportion of buyers in such a time period is given by

$$b_T = 1 - (1 + \frac{Tm}{k})^{-k}$$

This quantity is often called the penetration. The longer the time period, the larger will be the penetration.

The LSD approximation to this formula can be easily obtained by noting that the value of a_L is proportional to the length of the time period. Then we find

$$b_T \,/\, b = \ln(1 + Ta_L) \,/\, \ln(1 + a_L)$$

In this application it is definitely preferable to use a_L rather than q.

Probably the most important use of the NBD/LSD model is in the study of repeat-buying behavior. This is concerned with purchasing behavior in two consecutive equal time periods. A "repeat" buyer is one who buys at least once in both time periods. A "lost" buyer buys in period I but not in period II. Conversely a "new" buyer buys in period II but not in period I. Note that these definitions do not account for whether or not purchases were made in earlier or later periods.

Under stationary conditions, the proportion of new buyers in the population, b_N, is equal to the proportion of lost buyers b_L. As the proportion of new buyers is the difference between the proportion of buyers in the combined period and the proportion of buyers in period I, we have

$$b_N = b_L = 1 - (1 + \tfrac{2m}{k})^{-k} - [1 - (1 + \tfrac{m}{k})^{-k}]$$

$$= (1 + \tfrac{m}{k})^{-k} - (1 + \tfrac{2m}{k})^{-k},$$

where the distribution of purchases in period I is a NBD with parameters m and k. The proportion of repeat buyers, b_R, is given by

$$b_R = b - b_N$$

$$= 1 - 2\,(1 + \tfrac{m}{k})^{-k} + (1 + \tfrac{2m}{k})^{-k}.$$

The amounts bought by the different types of buyer can be obtained as follows. Since the purchases of a particular consumer in successive time periods are independent Poisson variates, a person whose average rate of buying is, say, λ, will buy λ units on average in period II regardless of how much he happened to buy in period I. Consider the non-buyers in period I. In period II the mean amount bought by this subgroup on a per-consumer basis is given by

$$\int_0^\infty f_p(\lambda)\, e^{-\lambda}\, \lambda\, d\lambda$$

$$= \frac{m}{(1 + a)^{k + 1}} = m_N,$$

since this must all be purchased by "new" buyers. Under stationary conditions this will be the same as the mean amount bought in period I by the lost buyers, namely m_L. (It is also equal to P(1)).

The mean amount bought by repeat buyers, m_R, is given by

$$m_R = m - m_N$$

$$= m[1 + (1 + a)^{-(k + 1)}].$$

The quantities m, m_N and m_R are calculated on a per-consumer basis. In practice it is often more useful to calculate quantities analagous to w, the rate of buying per buyer. The average rate of buying per new, lost and repeat buyer can be obtained from the formulae

$$w_N = w_L = m_N/b_N.$$

$$w_R = m_R/b_R.$$

It is a simple matter to find the LSD approximations to these formulae. For example, we find

$$b_R/b = 1 + \frac{\ln(1 + q)}{\ln(1 - q)} ;$$

$$m_R/m = q ;$$

$$w_R = - q^2/(1 - q)\ln(1 - q^2).$$

The LSD formulae have several advantages over their NBD counterparts. First, they depend only on one parameter, q. Second, since q is a function of w, they emphasize that repeat-buying behavior depends mainly on the average rate of buying per buyer. Third, since the formulae can be evaluated rapidly for different values of w (and hence of q), it is possible to simplify some of the formulae even further. For example, we find $w_R/w \simeq 1.2$ and $w_N = w_L \simeq 1.4$.

The differences between the NBD and LSD predictions depend on the values of w and b, but are generally fairly small. For example, the average rate of buying per repeat buyer was calculated by both methods, and the ratio of the results is given in Table 3.

Table 3. Values of w_R(LSD)/w_R(NRD) for different values of w and b.

		b 0.1	0.2	0.3	0.4
	1.6	1.022	1.046	1.074	1.108
	2.4	1.015	1.033	1.052	1.074
w	3.2	1.103	1.028	1.044	1.063
	4.0	1.012	1.026	1.040	1.057
	7.0	1.010	1.021	1.034	1.047
	10.0	1.009	1.019	1.030	1.043
	15.0	1.008	1.017	1.027	1.039

Similarly the proportion of repeat buyers was calculated by both methods and the ratio of the results is given in Table 4.

Table 4. The ratio b_R(LSD)/b_R(NBD) for different values of w and b

		b 0.1	0.2	0.3	0.4
	1.6	0.970	0.937	0.901	0.861
	2.4	0.979	0.957	0.933	0.906
	3.2	0.983	0.965	0.945	0.923
w	4.0	0.985	0.969	0.952	0.933
	7.0	0.988	0.976	0.963	0.948
	10.0	0.990	0.979	0.967	0.954
	15.0	0.991	0.982	0.971	0.960

The worst discrepancies occur when b is large and w is small. Generally speaking, the difference in predictions is less than 5% provided that $b < 0.075 + w/16$. However, if $b > 0.075 + w/16$ (which is rather unusual), then it may be advisable to use the NBD formulae rather than the LSD approximations.

The repeat-buying predictions have been found to hold under a wide range of conditions (Chatfield et al. [2]). The use of these predictions has been discussed by Ehrenberg [7]. Not only can they be used to analyze data in stationary situations, but they can also be used as a yardstick in nonstationary situations.

It is of interest to note that the repeat-buying formulae can also be obtained by using the fact that purchases by the same sample in different time periods will follow a multivariate NBD. Let m and k be the NBD parameters in some basic time period. If there are t periods and the ith period is T_i times the length of the basic time period, then the probability-generating function of the multivariate distribution of purchases in the different time periods is given by

$$1 + \frac{m}{k} \sum_{i=1}^{t} T_i \, (1 - u_i)^{-k}$$

A discussion of the properties and usefulness of the multivariate NBD (and multivariate LSD) is given by Chatfield et al. [2] together with further references.

4. THE BBD MODEL

The NBD/LSD model has given a variety of useful results. Nevertheless, certain systematic discrepancies from this model were noted, particularly the variance discrepancy. This discrepancy seems to occur for certain heavily bought products where a substantial number of consumers buy at least one unit of the product regularly every week. In order to 'explain' these discrepancies, a completely new study of consumer purchasing behavior was made.

Another way of presenting the data obtained from a consumer panel is to construct what is called the frequency distribution of "weeks." Given n weeks of data for a particular brand (or a specific pack-size), this distribution is given by $f_o, f_1, \ldots, f_n$, where

f_r = number of consumers in the sample who bought at least one unit of the brand in exactly r out of n weeks.

Two examples of distributions of this type are given in Table 5. Note particularly that they are truncated at the number of weeks in the time period. Chatfield and Goodhardt [3] noted that the beta-binomial distribution gives a good fit to distributions of weeks. This suggested the following compound binomial model:

(a) The probability that a given consumer will buy at least one unit of the brand (or pack-size) in a particular week is a constant p, which is independent of previous purchases. Thus, in a time period of n weeks, the number of weeks in which the consumer buys at least one unit will follow a binomial distribution with parameters n and p.

(b) The probability p varies from consumer to consumer, the distribution being a beta distribution.

Table 5. Examples of the fit of the BBD.

Number of weeks	Brand M (12 Weeks) Observed Frequency	BBD	Brand D (4 Weeks) Observed Frequency	BBD
0	25	25.0	410	410.0
1	7	8.5	22	22.5
2	6	5.1	16	13.8
3	4	3.5	9	11.9
4	3	2.5	17	15.8
5	2	1.8		
6	1	1.3		
7	1	0.9		
8	0	0.6		
9	1	0.4		
10	0	0.2		
11	0	0.1		
12	0	0.0		

Brand M: $s^2 = 4.46$, $\alpha = 0.390$, $\beta = 2.730$, $\sigma^2 = 4.82$

Brand D: $s^2 = 0.83$, $\alpha = 0.049$, $\beta = 0.576$, $\sigma^2 = 0.82$

This model is somewhat similar to that proposed by Pyatt [10], but it is concerned with 'weeks' rather than purchases and seems a more realistic approach.

An immediate consequence of this compound binomial model is that the distribution of weeks will follow a beta-binomial distribution. The p.d.f. of the beta distribution is given by

$$f_B(p) = \frac{p^{\alpha-1}(1-p)^{\beta-1}}{B(\alpha,\beta)}, \qquad 0 \le p \le 1, \quad \alpha > 0;\ \beta > 0.$$

where $B(\alpha,\beta) = \int_0^1 p^{\alpha-1}(1-p)^{\beta-1}\,dp$.

Then the proportion of the population who purchase at least one unit of the product in exactly r out of n weeks is given by

$$P(r) = \int_0^1 f_B(p)\binom{n}{r}p^r(1-p)^{n-r}dp$$

$$= \binom{n}{r} B(\alpha+r,\ n+\beta-r)/B(\alpha,\beta) \qquad r = 0,1,\ldots,n.$$

This distribution is also called the compound binomial, the negative hypergeometric and the Polya distribution. Its mean and variance are given by

$$\mu = n\alpha/(\alpha+\beta)$$

$$\sigma^2 = n\alpha\beta\ (n+\alpha+\beta)/(\alpha+\beta)^2(1+\alpha+\beta).$$

The shape of the distribution depends on the values of the two parameters α and β. In this investigation the distribution always has a mode at zero in which case α is less than one. In addition there is often a second mode at n in which case β is also less than one. The distribution is then called U-shaped.

As for the NBD, the standard method of fitting the BBD to consumer purchasing distributions is by mean and zeros. An iterative method for doing this is given by Chatfield and Goodhardt [3]. About 30 distributions of weeks for several different product fields were fitted in this way and generally speaking a good fit was obtained (see, for example, Table 5). In particular, no systematic variance discrepancy was observed.

Generally speaking, the distributions were of two types. First, those for which the parameter β was "small": If $\beta < 1 - \frac{1-\alpha}{n}$ the distribution is U-shaped; but even when β is somewhat larger, P(n) will still be substantial; and in both cases the corresponding distribution of purchases (or purchasing occasions) will give a variance discrepancy when fitted with a NBD.

The second type of distribution is that for which P(n) is "small." Provided that β is reasonably large (greater than about 2), the NBD will not give a variance discrepancy when fitted to the corresponding distribution of purchases. In fact, Skellam [11] has shown that the NBD is a limiting case of the BBD obtained by letting $\beta \to \infty$ in a certain way. If β and n are reasonably large, similar results will be obtained by fitting a NBD or BBD to the same distribution. An example is given in Table 6.

The importance of this is that for lightly bought products the distribution of purchases (or purchasing occasions) and weeks will be virtually identical, and both the NBD and BBD models are applicable. But for heavily bought products some consumers will purchase one or more units a week regularly. In such a case a variance discrepancy may result from fitting a NBD to the distribution of purchases, but the BBD will still describe the distribution of weeks with a small value of β.

Table 6. An example of the similarity between the NBD and BBD for the distribution of purchases for brand D over 12 weeks

Number of Units	Observed	NBD	BBD
0	875	875.0	875.2
1	63	59.1	56.4
2	19	22.1	22.9
3	10	10.6	11.6
4	4	5.6	6.4
5	4	3.1	3.6
6	1	1.8	2.0
7	2	1.1	1.1
8	0	0.7	0.5
9	1	0.4	0.3
10	0	0.2	0.1
11	0	0.2	0.0
12	1	0.1	0.0
	$s^2 = 0.72$	$m = 0.212$	$\alpha = 0.084$
		$a = 2.14$	$\beta = 4.670$
		$\sigma^2 = 0.67$	$\sigma^2 = 0.61$

This suggests that the BBD model is the more fundamental of the models. While it is reasonable to hypothesize that purchases in successive weeks are independent (as in the BBD model), the NBD model requires strictly that purchases in any two successive time periods will be independent, which is not the case.

From a practical point of view, the market researcher is more interested in purchases or purchasing occasions than weeks, and this focuses attention on the NBD model rather than the BBD model. Despite the variance discrepancy, the NBD model enables many useful predictions to be made. The BBD model is important because it throws light on the reasons for the NBD variance discrepancy and because it provides an accurate picture of one aspect of consumer purchasing for all types of consumer products.

5. THE BETA-GEOMETRIC DISTRIBUTION

An interesting consequence of the BBD model can be obtained by considering consumer purchasing in terms of the time between purchases rather than amounts bought. Consider a person whose probability of buying a particular brand in one week is equal to p. If this is independent of previous purchases, the probability that he makes his first purchase in the kth week is given by the geometric probabilities

$$P(k) = (1 - p)^{k-1}p \qquad k = 1,2,\ldots.$$

If p varies from consumer to consumer, the distribution being a beta distribution, then the proportion of consumers in the population who make their first purchase in the kth week is given by

$$P(k) = \int_0^1 f_B(p)(1 - p)^{k-1}\, p\, dp$$

$$= \frac{B(\alpha + 1, \beta + k - 1)}{B(\alpha,\beta)}, \qquad k = 1,2,\ldots,\infty$$

This distribution is called the beta-geometric, beta-Pascal, or negative infinite hypergeometric distribution. The rth factorial moment of the geometric distribution is equal to $r!(1-p)^{r-1}/p^r$, so that the rth factorial moment of the beta-geometric distribution is equal to

$$= \int_0^1 f_B(p)\, \frac{r!(1-p)^{r-1}}{p^r} dp = r!\frac{B(\alpha - r, \beta + r - 1)}{B(\alpha,\beta)},$$

provided that $\alpha > r$. But the distributions of weeks studied thus far give $0 < \alpha < 1$, so that the mean of the corresponding beta-geometric distribution will not then exist. For this reason, if data over n weeks is available, it is better to consider the distribution

$$P(k) = \frac{B(\alpha + 1, \beta + k - 1)}{B(\alpha,\beta)} \qquad k = 1,2,\ldots,n$$

$$P(n + 1) = \frac{B(\alpha, n + \beta)}{B(\alpha,\beta)}$$

$$= \text{proportion of non-buyers over } n \text{ weeks.}$$

An example of such a distribution is given in Table 7. The values of α and β were estimated from the corresponding distribution of weeks. The good fit of the distribution is further confirmation of the BBD model.

6. POSTSCRIPT

The NBD and BBD models are restricted to one brand at a time and no account is taken of whether or not a consumer has bought

Table 7. An example of the distribution of times to first purchase for brand M with data from 48 weeks.

Number of weeks	Observed Frequency	Beta-geometric
1	5	5.3
2	6	3.9
3 - 4	4	5.4
5 - 6	4	3.7
7 - 8	5	2.7
9 - 10	0	2.1
11 - 15	2	3.7
16 - 20	2	2.5
21 - 30	3	3.2
31 - 40	3	2.0
41 - 48	2	1.2
Did not buy	14	14.1
		α = 0.485
		β = 4.081

one or more other brands in the same product field.* In spite of this restriction, a large number of useful results have been obtained leading to an improved understanding of consumer purchasing behavior. Such work is also an essential preliminary to considering more complex situations. Ideally, one would like a model which would describe brand-switching, multiple purchases of two or more brands, and behavior in nonstationary conditions. Markov chain models have focused attention on brand-switching, but these models have several disadvantages (Ehrenberg, [6]) which have not yet been overcome (and are not likely to be). A more fruitful approach would appear to be in terms of some multivariate distribution.

REFERENCES

[1] Anscombe, F. J. 1950. Sampling theory of the negative binomial and logarithmic distributions. Biometrika, 37:358-82.

[2] Chatfield, C; Ehrenberg, A.S.C.; and Goodhardt, G. J. 1966. Progress on a simplified model of stationary purchasing behavior. J. Roy. Statist. Soc. Sec. A., 129:317-67.

* (The BBD model has also been successfully used in a few cases to analyze data obtained from all purchases in a product field).

[3] Chatfield, C., and Goodhardt, G. J. 1969. The beta-binomial model for consumer purchasing behavior. (To be published).

[4] Chatfield, C. 1969. On estimating the parameters of the logarithmic series and negative binomial distributions. Biometrika, 56.

[5] Ehrenberg, A. S. C. 1959. The pattern of consumer purchases. Appl. Statist. 8:26-41.

[6] ———, 1965. An appraisal of Markov brand-switching models. J. Marketing Research 2:347-62.

[7] ———, 1968. The practical meaning and usefulness of the NBD/LSD theory of repeat-buying. Appl. Statist., 17:17-32.

[8] Fisher, R. A.; Corbet, A. S.,; and Williams, C. B. 1943. The relation between the number of species and the number of individuals in a random sample of an animal population. J. Animal Ecology, 12:42-58.

[9] Patil, G. P.; Kamat, A. R.; and Wani, J. K. 1964. Certain Studies on the Structure of the Logarithmic Series Distribution and Related Tables. Ohio: Aerospace Research Laboratories.

[10] Pyatt, G. 1969. A model of brand loyalties. (In the press). Proc. 1968 CEIR/SCS Symposium on Model-Building in Business and Industry. Griffin.

[11] Skellam, J. G. 1948. A probability distribution derived from the binomial distribution by regarding the probability of success as variable between the sets of trials. J. Roy. Statist. Soc. Ser. B., 10:257-61.

[12] Williamson, E., and Bretherton, M. H. 1964. Tables of the logarithmic series distribution. Ann. Math. Statist., 35:284-97.

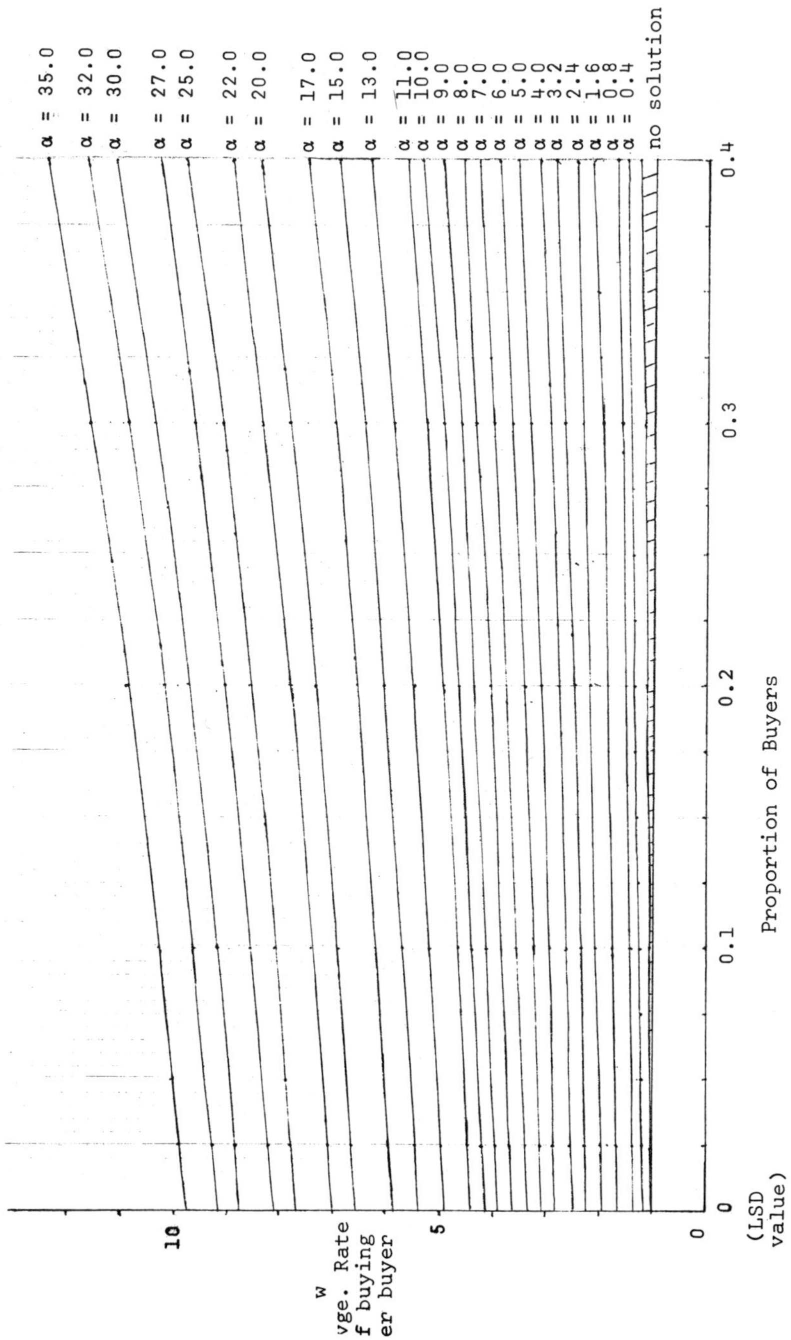

Fig. 1. Estimating the NBD Parameter A by Mean and Zeros

CHANCE OF ALL SUCCESSES FOR INDEPENDENT BINOMIAL EVENTS WITH LARGE SUCCESS PROBABILITIES

JOHN E. WALSH and GRACE J. KELLEHER
Southern Methodist University*

SUMMARY

Consider reliability situations where all of n independent steps or operations must be satisfactory (successes) if the overall operation is to be satisfactory. The interest is in reliability levels that are at least moderately high. That is, the probability of obtaining successes for all steps or operations is moderately near unity. An approximate expression, also sharp upper and lower bounds , are developed for the probability of all successes. This is done for a range of levels of generality, including complete generality. These results depend only on the generality level, the number of steps or operations, and the arithmetic average of the probabilities of success for the steps and operations. They are useful when the probability of all successes is at least 0.75; then, the approximate value is moderately near both bounds. The relative error of the approximate expression is at most one percent when its value is at least 0.83. Statistical estimation of the probability of all successes is simplified when an expression using the arithmetic average of the success probabilities is used.

*Research partially supported by Mobil Research and Development Corp., ONR Contract N00014-68-A-0515, and NASA Grant NGR 44-007-028.

I. INTRODUCTION AND RESULTS

Consider n independent binomial events, where success and failure are the possible outcomes for an event. There is sometimes interest in whether all the events are successes, or whether all the events are failures. As an example, some reliability situations are such that overall operation is satisfactory if and only if all of n independent steps or operations are satisfactory (successes). As another example, there are situations where occurrence of at least one success is enough to assure the overall success of a project (for example, in research efforts). Of course, the probability of at least one success is unity minus the probability of all failures.

High reliabilities (say, probability of overall success is at least .8) are often of interest. Also, research projects where the probability of at least one success is too small (say $\leq p$, where $p \leq 0.2$) probably should not be undertaken. These are the ranges of probability levels for which approximate values and bounds are developed; that is, the principal interest is in cases where the probability of all successes is at least 0.8, or cases where the probability of all failures is at least 0.8. Only the probability of all successes is considered in the remaining material. However, corresponding results are obtained for all failures by exchanging the roles of success and failure.

Let p_i be the probability of success for the ith of the n independent steps or operations. The arithmetic average of the p_i is denoted by p. The level of generality is defined in terms of the number r of the p_i that are required to be p or less in value. Complete generality occurs for r = 1, and the generality level decreases as r increases.

Only n, p, and r such that $(n + 1 - r)(1 - p) < 1$ receive consideration. Then, for any meaningful n and r that are at least unity,

$$[1 - (n + 1 - r)(1 - p)]p^{r-1} \leq P(\text{all successes}) \leq p^n .$$

The bounds are close together and P(all successes) is very nearly equal to

$$(1/2)p^{r-1}[p^{n+1-r} + 1 - (n + 1 - r)(1 - p)]$$

when $p \geq 1 - .2/(n + 1 - r)$.

It is to be noted that exact values for P(all successes)

could be obtained in terms of the geometric mean of the success probabilities of the individual events. That is, P(all successes) equals this geometric mean raised to the nth power. However, the multiplicative character of the geometric mean makes it substantially less convenient for use than p (with respect to estimation of its value, etc.) except when it can be approximately expressed in terms of p. For example, the observed number of successes for the n events in an unbiased estimate of p. Thus, past data on similar sets of events can be directly used to estimate p.

The next and final section contains derivations for the results stated above. The results for r = 1 could also have been developed from the material of (Hoeffding, 1956).

2. DERIVATIONS

Since the geometric mean of the p_i is at most equal to the arithmetic mean, P(all successes) is at most equal to p^n, with equality possible when all the p_i are equal (holds for all r).

Now, consider derivation of the sharp lower bound. The value of P(all successes) is

$$\prod_{i=1}^{n} p_i = \exp\left\{\sum_{i=1}^{n} \log_e[1 - (1 - p_i)]\right\}$$

$$= \exp\left[-\sum_{i=1}^{n}\sum_{j=1}^{\infty} (1 - p_i)^j/j\right]$$

$$= \exp\left[-\sum_{i=1}^{\infty} j^{-1} \sum_{k=0}^{j} \binom{j}{k} (1 - p_i)^{j-k} \sum_{i=1}^{n} (p - p_i)^k\right]$$

For $(n + 1 - r)(1 - p) < 1$ and $k \geq 2$, $\sum_{i=1}^{n} (p - p_i)^k$ is largest when all but r of the p_i are unity, r - 1 of them equal p, and the remaining one is such that their arithmetic average is p. This is easily verified by considering the case of k = 2,3 and the relative relationships for larger k. This implies that the remaining p_i equals $1 - (n + 1 - r)(1 - p)$. For k=0 and 1,

the values of $\sum_{i=1}^{n} (p - p_i)^k$ are all the same for all possible selections of the p_i. Thus, the effect of maximizing $\sum_{i=1}^{n} (p - p_i)^k$ for $k \geq 2$ is to minimize the argument of the exponential. This implies that the lower bound for P(all successes) is

$$[1 - (n + 1 - r)(1 - p)]p^2 ,$$

the value that occurs for the selection to the p_i so that $\sum_{i=1}^{n} (p - p_i)^k$ is maximized for $k \geq 2$.

REFERENCE

Hoeffding, Wassily, 1956. "On the distribution of the number of successes in independent trials," Annals of Math. Stat., 27:713-21.

A COMPOUND POISSON PROCESS MODEL FOR MULTIPLE SECURITY ANALYSIS

S. JAMES PRESS
University of Chicago

SUMMARY

A statistical model based upon the compound Poisson process is proposed for studying the probability distribution of price changes of many stock market securities simultaneously. This work extends previously proposed security price behavior models involving the Compound Poisson process to the multivariate case. Results analogous to the univariate case are obtained. The model permits price movements of different securities to be correlated with one another. It is shown that the distributions of price changes in this model have precisely the properties which have been found empirically; that is, empirical distributions and those in this model are generally skew-symmetric, peaked "fat-tailed," and, in the absence of a trend, unimodal. A moment-matching method is suggested for estimating the parameters.

1. INTRODUCTION

Price fluctuations of securities in the stock market have been analyzed both statistically and intuitively. A summary of some of the investigations carried out prior to 1964 is conveniently provided in a book edited by Cootner [1]. Recently, statistical research in this area has been focused on the empirically observed fact that successive changes in price of a particular stock are approximately uncorrelated (random walk model) and seem to follow a probability distribution which assigns appreciable probability to large deviations of price changes from the mean value, is usually skew-symmetric and unimodal, and is more highly peaked around its mean than a comparable normal random variable. Models which account for such random walk nonnormal distributions include the stable Paretian model (see Mandelbrot, [5]; Fama, [3]), in which the distributions contain no moments, and the Compound Poisson Model of Press [8], which has the advantage that all moments exist. In those models, price changes of a single security are shown to follow a distribution which possesses the above-mentioned properties.

In this paper, we extend the Compound Poisson Process Model to the multivariate case by suggesting a model which explains simultaneous behavior of many securities, permitting the price changes of the various securities to be mutually correlated. It is shown that, while separate interrelated "market factors" and "industry factors" are permitted, the model still retains the property of the univariate model that each of the stocks separately possesses the required type of distribution for price fluctuations. For additional background and motivation for this type of a multivariate model, see King [4] and Press [7].

In the next Section we present and discuss the basic model used for multiple security analysis. The properties of the model are developed in Section 3, and in Section 4 the model parameters are estimated.

2. MODEL

Let $Z_j(t)$ denote the natural logarithm of the price of security j at time t, $j = 1,\ldots,p$, and let $Z(t) \equiv [Z_1(t),\ldots,Z_p(t)]$ denote the row vector of prices for all securities under consideration. Denote the initial values of the "prices" (logged prices) by $C \equiv [Z_1(0),\ldots,Z_p(0)] \equiv [C_1,\ldots,C_p]$, where C is assumed to be known and fixed. Now suppose that the mechanism for generating $Z_j(t)$ is that "price" at any time is composed of an initial "price." a sum of changes in price from its initial value peculiar to the jth security, a sum of price changes correlated with the movements of the market (or group) as a whole, and finally, some otherwise unexplainable "noise" manifested as random price fluctuations representative of Brownian movement.

The defining vector equation is

$$Z(t) = C + H(t) + \sum_{k=1}^{M(t)} W_k + X(t), \tag{2.1}$$

where $H(t)$, W_k, and $X(t)$ are the p-variate vectors.

$$H(t) \equiv \left[\sum_{k=1}^{N_1(t)} Y_{k1},\ldots, \sum_{k=1}^{N_p(t)} Y_{kp}\right], \tag{2.2}$$

$$W_k \equiv [W_{k1},\ldots,W_{kp}],$$

$$X(t) \equiv [X_1(t),\ldots,X_p(t)].$$

Each component of $H(t)$ is a compound Poisson process with normal compounding; that is, for the jth security,

$$\mathcal{L}(Y_{kj}) = N(\theta_j,\ \sigma_j^2), \quad \text{for all } k, \tag{2.3}$$

and $N_j(t)$ is a Poisson process with mean $\lambda_j t$ which represents the number of random events (not necessarily transactions) which are accompanied by price changes; $\{N_j(t),\ t \geq 0\}$ is assumed to be independent of the Y_{kj}'s, and the components of $H(t)$ are assumed to be mutually independent so that they account for price changes specific to each of the separate securities being studied.

$\{M(t),\ t \geq 0\}$ is a Poisson process which represents the number of price-changing events during $[0,t]$ which affect all securities in the group; its mean is γt. Assume that the p-vector W_k follows the distribution

$$\mathcal{L}(W_k) = N(\mu,\Sigma),\ \Sigma > 0, \text{ for every } k, \tag{2.4}$$

and that the W_k's are mutually independent and also independent of $\{M(t),\ t \geq 0\}$. Thus $\Sigma_1^{M(t)} W_k$ is a vector of compound Poisson processes, and the components of the vector are correlated. Hence this term in the model accounts for price changes and price-changing events which affect all securities in the group in similar though not necessarily identical ways.

$\{X_j(t),\ t \geq 0\}$ denotes the Wiener Process associated with the jth security. Accordingly, assume

$$\mathcal{L}\{X_j(t)\} = N(0,\ t\sigma^2),\ \text{for all } j, \tag{2.5}$$

and that the components of X(t) are mutually independent.

Finally, assume H(t), $\Sigma_1^{M(t)} W_k$ and X(t) are mutually independent.

We note in passing that by setting $\gamma = 0$, we obtain the single security or univariate model which was studied earlier.

3. DISTRIBUTIONAL PROPERTIES OF THE MODEL

In addition to the model given in (2.1) we will be interested in studying the increments model

$$\Delta Z(t) \equiv Z(t) - Z(t-1) = \Delta H(t) + \sum_{k=M(t-1)+1}^{M(t)} W_k + \Delta X(t), \tag{3.1}$$

where $\Delta X(t) = X(t) - X(t-1)$, and

$$\Delta H(t) = H(t) - H(t-1) \equiv \left[\sum_{k=N_1(t-1)+1}^{N_1(t)} Y_{k1}, \ldots, \sum_{k=N_p(t-1)+1}^{N_p(t)} Y_{kp} \right].$$

Theorem 1:

$\{Z(t),\ t \geq 0\}$ is a vector process of stationary and independent increments.

Proof: The jth component of Z(t) is clearly expressible as

$$Z_j(t) = C_j + \sum_{k=1}^{N_j(t)} Y_{kj} + \sum_{k=1}^{M(t)} W_{kj} + X_j(t),\ j = 1,\ldots,p. \tag{3.2}$$

Since each of the processes in (3.2) has stationary and independent increments (see Parzen [6], pp. 128-31), so does the sum.

Theorem 2:

$$E[Z(t)] = C + t[\theta_1\lambda_1 + \gamma\mu_1,\ldots,\theta_p\lambda_p + \gamma\mu_p], \tag{3.3}$$

and

$$E[\Delta Z(t)] = [\theta_1\lambda_1 + \gamma\mu_1,\ldots,\theta_p\lambda_p + \gamma\mu_p], \tag{3.4}$$

where $\mu \equiv [\mu_1,\ldots,\mu_p]$.

Proof: Take expectations in (2.1) and (3.1) by conditioning on $[N_1(t),\ldots,N_p(t); M(t)]$ and $[N_1(t)-N_1(t-1,\ldots,N_p(t)-N_p(t-1); M(t)-M(t-1)]$, respectively.

Theorem 3:

The covariance matrices of $Z(t)$ and $\Delta Z(t)$ are given, respectively, by

$$\Psi(t) = \text{diag}[\sigma^2 t+\lambda_1 t(\theta_1^2+\sigma_1^2),\ldots,\sigma^2 t+\lambda_p t(\theta_p^2+\sigma_p^2)] + \gamma t(\Sigma + \mu'\mu). \quad (3.5)$$

$$\Delta\Psi(t) = \text{diag}[\theta^2+\lambda_1(\theta_1^2+\sigma_1^2),\ldots,\sigma^2+\lambda_p(\theta_p^2+\sigma_p^2] + \gamma(\Sigma + \mu'\mu). \quad (3.6)$$

Proof: Denote the covariance matrices of $H(t)$, $\sum_{k=1}^{M(t)} W_k$, and $X(t)$ by $\Psi_1(t)$, $\Psi_2(t)$, and $\Psi_3(t)$, respectively. Then by independence,

$$\Psi(t) = \Psi_1(t) + \Psi_2(t) + \Psi_3(t).$$

Since the components of $H(t)$ are independent,

$$\Psi_1(t) = \text{diag}\left[\text{Var}\left(\sum_{k=1}^{N_1(t)}\right),\ldots, \text{Var}\left(\sum_{k=1}^{N_p(t)} Y_{kp}\right)\right]$$

But by conditioning on $N_j(t)$ it is easily seen that

$$\text{Var}\left(\sum_{k=1}^{N_j(t)}\right)Y_{kj} = \lambda_j t(\theta_j^2 + \sigma_j^2), \quad (3.7)$$

which gives $\Psi_1(t)$.

$\Psi_2(t)$ is also found by conditioning

$$\Psi_2(t) = \text{Var}(\sum_{k=1}^{M(t)} W_k) = E(\sum_{j=1}^{M(t)} W_j')(\sum_{k=1}^{M(t)} W_k) - (\gamma t\mu')(\gamma t\mu),$$

where $\gamma t\mu' = E(\sum_{j=1}^{M(t)} W_j')$. Then since

$$E(\sum_{j=1}^{M(t)} W_j')(\sum_{k=1}^{M(t)} W_k) = E\left\{E(\sum_{j=1}^{m} W_j')(\sum_{k=1}^{m} W_k)\,|\,M(t) = m\right\}$$

$$= E\left\{E(\sum_{j=1}^{m} W_j'W_j)\,|\,M(t) = m\right\} + E\left\{E(\sum_{j=1}^{m}\sum_{\substack{k=1\\ j\neq k}}^{m} W_j'W_k)\,|\,M(t) = m\right\}$$

$$= E\{M(t)(\Sigma + \mu'\mu)\} + E\{[M^2(t) - M(t)]\mu'\mu\},$$

$$\Psi_2(t) = \gamma t(\Sigma + \mu'\mu), \tag{3.8}$$

from (2.5),

$$\text{Var}\,[X(t)] = \sigma^2 t\, I_p, \tag{3.9}$$

where I_p denotes the identity matrix of order p.

Combining (3.7), (3.8), and (3.9) gives (3.5). The demonstration of (3.6) is completely analogous.

Theorem 4:

$\{Z(t),\ t \geq 0\}$ and $\{\Delta Z(t),\ t \geq 0\}$ have log characteristic functions given by

$$\ln \phi(s) = isC' + \sum_{k=1}^{p} \lambda_k t \left[e^{i\theta_k s_k - \frac{1}{2}\sigma_k^2 s_k^2} - 1 \right] - \frac{1}{2}\sigma^2 tss' + \gamma t \left[e^{is\mu' - \frac{1}{2}s\Sigma s'} - 1 \right] \tag{3.10}$$

$$\ln \Delta\phi(s) = \sum_{k=1}^{p} \lambda_k \left[e^{i\theta_k s_k - \frac{1}{2}\sigma_k^2 s_k^2} - 1 \right] - \frac{1}{2}\sigma^2 ss' + \gamma \left[e^{is\mu' - \frac{1}{2}s\Sigma s'} - 1 \right], \tag{3.11}$$

respectively, where $s \equiv (s_1, \ldots, s_p)$ denotes the p-vector of characteristic function variables.

Proof: From (2.1),

$$\phi(s) = Ee^{is[C' + H'(t) + \Sigma_1^{M(t)} W_k' + X'(t)]},$$

and by independence,

$$\phi(s) = e^{isC'} \cdot Ee^{isH'(t)} \cdot Ee^{is\Sigma_1^{M(t)} W_k'} \cdot Ee^{isX'(t)}.$$

Using the characteristic functions of a multivariate normal distribution and that of a compound Poisson process (see, e.g., Parzen [6], pp. 128-31) and taking logs gives (3.10); (3.11) is obtained analogously using (3.1).

Theorem 5:

The density of $\Delta Z(t)$ is given by

$$f(x) = \sum_{n_1=0}^{\infty} \cdots \sum_{n_p=0}^{\infty} \sum_{m=0}^{\infty} \left(\frac{e^{-\lambda_1}\lambda_1^{n_1}}{n_1!}\right) \tag{3.12}$$

$$\cdots \left(\frac{e^{-\lambda_p}\lambda_p^{n_p}}{n_p!}\right)\left(\frac{e^{-\gamma}\gamma^m}{m!}\right) \cdot \frac{1}{(2\pi)^{\frac{p}{2}}|\Omega|^{\frac{1}{2}}} e^{-\frac{1}{2}(x-\phi)'\Omega^{-1}(x-\phi)},$$

where $x \equiv (x_1,\ldots,x_p)$, $\phi \equiv m\mu + (n_1\theta_1,\ldots,n_p\theta_p)$, and $\Omega = m\Sigma + \mathrm{diag}(\sigma^2 + n_1\sigma_1^2,\ldots,\sigma^2 + n_p\sigma_p^2)$.

Proof: From (2.5) and the fact that X(t) is a Wiener process,

$$\mathcal{L}\{\Delta X(t)\} = N(0,\sigma^2 I_p). \tag{3.13}$$

From (2.4) it is clear that, by conditioning, we obtain

$$\mathcal{L}\left\{\sum_{k=M(t+1)+1}^{M(t)} W_k \,\middle|\, M(t) - M(t-1) = m\right\} = N(m\mu,\ m\Sigma). \tag{3.14}$$

Finally, using (2.3) and conditioning, we obtain

$$\mathcal{L}\left\{\Delta H(t) \,\middle|\, N_1(t) - N_1(t-1) = n_1,\ldots,N_p(t) - N_p(t-1) = n_p\right\} \tag{3.15}$$

$$= N\left[(n_1\theta_1,\ldots,n_p\theta_p),\ \mathrm{diag}(n_1\sigma_1^2,\ldots,n_p\sigma_p^2)\right].$$

Substituting (3.13)-(3.15) in (3.1) and conditioning gives

$$\mathcal{L}\left\{\Delta Z(t) \,\middle|\, N_1(t)-N_1(t-1) = n_1,\ldots,N_p(t)-N_p(t-1) = n_p,\ M(t)-M(t-1) = m\right\}$$

$$= N\left[m\mu + (n_1\theta_1,\ldots,n_p\theta_p),\ \sigma^2 I_p + m\Sigma + \mathrm{diag}(n_1\sigma_1^2,\ldots,n_p\sigma_p^2)\right] \tag{3.16}$$

$$= N(\phi,\Omega).$$

Hence, unconditioning gives (3.12).

Note that the series (3.12) converges for all x on $(-\infty,\infty)$, being dominated uniformly in x by absolutely convergent series of Poisson coefficients.

Corollary 5.1:

The marginal density of $\Delta Z_j(t)$ is given by

$$f_j(x_j) = \sum_{n_j=0}^{\infty} \sum_{m=0}^{\infty} \left(\frac{e^{-\lambda_j}\lambda_j^{n_j}}{n_j!}\right)\left(\frac{e^{-\gamma}\gamma^m}{m!}\right) \frac{\exp\left\{-\frac{1}{2\omega_{jj}}(x_j - \phi_j)^2\right\}}{\omega_{jj}^{\frac{1}{2}}\sqrt{2\pi}}, \tag{3.17}$$

where $\phi_j = (m\mu_j + n_j\theta_j)$ is the jth component of ϕ,

$\omega_{jj} \equiv \sigma^2 + n_j\sigma_j^2 + m\sigma_{jj}$, and $\Sigma \equiv (\sigma_{ij})$, i, j = 1,...,p.

Proof: From (3.16),

$$\mathscr{L}\left\{\Delta Z_j(t) \middle| N_1(t) - N_1(t-1) = n_1,\ldots,N_p(t) - N_p(t-1) = n_p, \right. \tag{3.18}$$

$$\left. M(t) - M(t-1) = m\right\}.$$

$$= N\left[m\mu_j + n_j\theta_j,\ \sigma^2 + m\sigma_{jj} + n_j\sigma_j^2\right] = N(\phi_j,\omega_{jj}).$$

Since ϕ_j and ω_{jj} depend on n_j and m only, after unconditioning all other sums are unity.

Theorem 6:

The cdf. of $\Delta Z(t)$ is given by

$$F(y) \equiv F(y_1,\ldots,y_p) = P\left\{\Delta Z_1(t) \leq y_1,\ldots,\Delta Z_p(t) \leq y_p\right\} \tag{3.19}$$

$$= \sum_{n_1=0}^{\infty}\cdots\sum_{n_p=0}^{\infty}\sum_{m=0}^{\infty}\left(\frac{e^{-\lambda_1}\lambda_1^{n_1}}{n_1!}\right)\cdots\left(\frac{e^{-\lambda_p}\lambda_p^{n_p}}{n_p!}\right)\left(\frac{e^{-\gamma}\gamma^m}{m!}\right)$$

$$\cdot\int_{-\infty}^{y_1}\cdots\int_{-\infty}^{y_p}\frac{1}{(2\pi)^{\frac{p}{2}}|\Omega|^{\frac{1}{2}}}e^{-\frac{1}{2}(x-\phi)'\Omega^{-1}(x-\phi)}dx.$$

Proof: Use (3.12) in the definition

$$F(y) = \int_{-\infty}^{y_1}\cdots\int_{-\infty}^{y_p} f(x)dx.$$

Theorem 7:

$\{\Delta Z(t),\ t \geq 0\}$ is a martingale if and only if $\theta \equiv (\theta_1,\ldots,\theta_p) = 0$ and $\mu = 0$.

Proof: $\{Z(t),\ t \geq 0\}$ is a vector-independent increments process. Hence, it is a martingale if $E[\Delta Z(t)] = 0$; i.e., if $\theta = \mu = 0$ (see Doob [2], pp. 96-7). The converse follows from the definition of a martingale.

Theorem 8:

The distributions of all components of $\Delta Z(t)$ are leptokurtic.

Proof: The kurtosis of a distribution is defined as $\gamma_{2j} \equiv K_{4j}/K_{2j}^2$, where $K_{\alpha j}$ is the αth cumulant for the jth component of $\Delta Z(t)$. If $\gamma_{2j}>0$, the distribution is leptokurtic.

Let all components of s, except the jth, vanish in (3.11). The result is the characteristic function of the jth component of $\Delta Z(t)$:

$$\ell n\ \Delta\phi(s_j) = -\frac{\sigma^2 s_j^2}{2} + \lambda_j\left[e^{i\theta_j s_j - \frac{1}{2}\sigma_j^2 s_j^2} - 1\right] \tag{3.20}$$

$$+ \gamma\left[e^{is_j\mu_j - \frac{1}{2}\sigma s_j^2\ jj} - 1\right].$$

If each of the exponential terms in this equation is expanded in a power series in s_j about $s_j = 0$, and the terms are combined, the result is expressible in the form

$$\ell n\ \Delta\phi(s_j) = \sum_{\alpha=1}^{4} \frac{K_{\alpha j}(is_j)^\alpha}{\alpha!} + o(s_j^4), \tag{3.21}$$

where the cumulants, $K_{\alpha j}$, are given by

$$K_{1j} = \theta_j\lambda_j + \gamma\mu_j,$$

$$K_{2j} = (\theta_j^2 + \sigma_j^2)\lambda_j + \gamma(\sigma_{jj} + \mu_j^2) + \sigma^2,$$

$$K_{3j} = 3\mu_j\sigma_{jj}\gamma + \mu_j^3\gamma + 3\theta_j\sigma_j^2\lambda_j + \theta_j^3\lambda_j,$$

$$K_{4j} = (3\sigma_{jj}^2 + 6\mu_j^2\sigma_{jj} + \mu_j^4)\gamma + (3\sigma_j^4 + 6\theta_j^2\sigma_j^2 + \theta_j^4)\lambda_j.$$

It is clear that since $K_{4j} > 0$, $\gamma_{2j} > 0$, for all $j = 1,\ldots,p$.

Theorem 9:

The distribution of $\Delta Z_j(t)$ for all j is more peaked in the vicinity of its mean than the distribution of a normally distributed random variable with the same mean and variance.

Proof: Define for fixed j,

$$V_j = \frac{\Delta Z_j(t) - E\Delta Z_j(t)}{\sqrt{\mathrm{Var}\ \Delta Z_j(t)}} = \frac{\Delta Z_j(t) - K_{1j}}{\sqrt{K_{2j}}}.$$

It suffices to show that the density of V_j in the vicinity of the origin exceeds the density of a standard normal variate.

Using (3.20) and the inversion formula gives for the density of V_j,

$$p(x_j) = \frac{1}{2\pi}\int_{-\infty}^{\infty} \exp\left\{- is_jx_j - \frac{s_j^2}{2} - \frac{is_j^3K_{3j}}{6K_{2j}^{\frac{3}{2}}} + \frac{K_{4j}s_j^4}{24K_{2j}^2} + o(s_j^4)\right\}ds_j.$$

Setting $x_j = 0$ and integrating termwise (since the integrand converges uniformly), gives

$$p(x_j) = \frac{1}{2\pi} \int_{-\infty}^{\infty} e^{-\frac{s_j^s}{2}} \left[1 - \frac{is_j^3 K_{3j}}{6K_{2j}^{\frac{3}{2}}} + \frac{K_{4j} s_j^4}{24K_{2j}^2} + o(s_j^4) \right] ds_j$$

$$= \frac{1}{\sqrt{2\pi}} + \frac{K_{4j}}{8K_{2j}^2 \sqrt{2\pi}} + \dots .$$

Note that $(2\pi)^{-\frac{1}{2}}$ is the density of a standard normal variable at the origin. Moreover, since all terms of the series obtained by integrating odd powers of s_j must vanish, all terms in the series are real (and will alternate in sign). Hence the magnitude of the remainder cannot exceed $K_{4j}\Big/\left(8K_{2j}^2\sqrt{2\pi}\right)$. But we saw in Theorem 8 that $K_{4j} > 0$.

Theorem 10:

The distribution of the vector $\Delta Z(t)$ is symmetric about its mean if and only if $E[\Delta Z(t)] = 0$.

Proof: It is clear from (3.4) that $E[\Delta Z(t)] = 0$ if and only if $\mu = 0$ and $\theta = 0$. Thus, if $E[\Delta Z(t)] = 0$ in (3.11), we obtain

$$\ln \Delta\phi(s) = -\frac{\sigma^2}{2} ss' + \sum_{k=1}^{p} \lambda_k \left(e^{-\frac{1}{2}\sigma_k^2 s_k^2} - 1\right) + \gamma\left(e^{-\frac{1}{2}s\Sigma s'}\right).$$

The right-hand side is real so that the distribution is symmetric.

Conversely, $\ln \Delta\phi(s)$ has an imaginary part as long as $\theta \neq 0$ or $\mu \neq 0$. So $E[\Delta Z(t)]$ must vanish in order for the distribution of $\Delta Z(t)$ to be symmetric.

Theorem 11:

The distribution of $\Delta Z(t)$ is unimodal if $\theta = \mu = 0$.

Proof: Set $\phi_j = 0$ in (3.17); differentiate with respect to x_j and set the result equal to zero. This gives

$$x_j \left[\sum_{n_j=0}^{\infty} \sum_{m=0}^{\infty} \left(\frac{e^{-\lambda_j} \lambda_j^{n_j}}{n_j!} \right) \left(\frac{e^{-\gamma}\gamma^m}{m!} \right) \frac{\exp\left\{-\frac{x_j^2}{2\omega_{jj}}\right\}}{\omega_{jj}^{\frac{3}{2}} \sqrt{2\pi}} \right] = 0.$$

Since the bracketed expression is positive for all x_j, $x_j = 0$ is a stationary point. Moreover, since $\partial^2 f_j(x_j)/\partial x_j^2$ is negative at $x_j = 0$, the origin is a unique mode; this holds for all

$j = 1,\ldots,p$. The identical argument holds, furthermore, for the multivariate density in (3.12).

Theorem 12:

For each component of $\Delta Z(t)$, the probability in the extreme tails of the distribution exceeds that of a normally distributed variable with the same mean and variance.

Proof: Let $r(x_j)$ denote the density of a normal random variable with the same mean and variance as $\Delta Z_j(t)$. Then if

$$\rho_j \equiv E[\Delta Z_j(t)] = \theta_j\lambda_j + \gamma\mu_j,$$

$$g_j^2 \equiv \mathrm{Var}[\ Z_j(t)] = \sigma^2 + \lambda(\theta_j^2 + \sigma_j^2) + \gamma(\sigma_{jj} + \mu_j^2),$$

$$r(x_j) = \frac{1}{g_j\sqrt{2\pi}} \exp\left\{-\frac{(x_j - \rho_j)^2}{2g_j^2}\right\}.$$

It must be shown that in the tails, $[f_j(x_j)/r(x_j)]$ exceeds unity. The ratio may be written as

$$\frac{f_j(x_j)}{r(x_j)} = \sum_{n_j=0}^{\infty}\sum_{m=0}^{\infty} \alpha(n_j,\ m)e^{Q(n_j,m)},$$

where $\alpha(n_j,m)$ is a constant not depending upon x_j, given by

$$\alpha(n_j,m) \equiv \left(\frac{e^{-\lambda_1}\lambda_1^{n_j}}{n_j!}\right)\left(\frac{e^{-\gamma}\gamma^m}{m!}\right)\left(\frac{g_j}{\omega_{jj}^{\frac{1}{2}}}\right)\exp\left\{\frac{1}{2}\left(\frac{\rho_j^2}{g_j^2} - \frac{\phi_j^2}{\omega_{jj}}\right)\right\},$$

and

$$Q(n_j,m) \equiv \frac{x_j^2}{2}\left(\frac{1}{g_j^2} - \frac{1}{\omega_{jj}}\right) - x_j\left(\frac{\rho_j}{g_j^2} - \frac{\phi_j}{\omega_{jj}}\right).$$

Since the sum of positive terms exceeds any one of its terms, for $n_j = \bar{n}_j$, and $m = \bar{m}$,

$$\frac{f_j(x_j)}{r(x_j)} > \alpha(\bar{n}_j,\bar{m})e^{Q(\bar{n}_j,\bar{m})}.$$

We can always choose $(\bar{n}_j,\bar{m})$ so large that $\omega_{jj} > g_j^2$. Then the coefficient of x_j^2 in $Q(n_j,m)$ is positive. Therefore, as $|x| \to \infty$, $Q \to \infty$. Hence, for all sufficiently large $|x|$, $[f_j(x_j)/r(x_j)] > 1$.

4. ESTIMATION

In order to carry out an empirical multiple security analysis using this model, it is necessary, of course, to estimate the parameters of the model. The estimation procedure developed below depends upon cumulant matching. The unknown parameters of the multiple security model are $(\mu; \Sigma; \gamma; \sigma^2; \theta; \lambda_1,\ldots,\lambda_p; \sigma_1^2,\ldots,\sigma_p^2)$. For estimation, assume that prices of all securities in the group are available for all times $t = 0,1,\ldots,T$. Equivalently, it is possible to form the observation vectors

$$[\Delta Z(1),\ldots,\Delta Z(T)],$$

which are independent and identically distributed. The sample mean vector and sample covariance matrix (unbiased) are

$$m_1 = \frac{1}{T}\sum_{k=1}^{T} \Delta Z(k),$$

and

$$S = \frac{1}{T-1}\sum_{k=1}^{T} [\Delta Z(k) - m_1]'[\Delta Z(k) - m_1],$$

respectively. Equating sample and population values gives

$$(\theta_1\lambda_1 + \gamma\mu_1,\ldots,\theta_p\lambda_p + \gamma\mu_p) = m, \tag{4.1}$$

$$\mathrm{diag}[\sigma^2 + \lambda_1(\theta_1^2 + \sigma_1^2),\ldots,\sigma^2 + \lambda_p(\theta_p^2 + \sigma_p^2)] + \gamma(\Sigma + \mu'\mu) = S. \tag{4.2}$$

Define the sample vectors

$$m_r = \frac{1}{T}\sum_{k=1}^{T} \Delta_r Z(k),$$

where $r = 1,\ldots,6$, and

$$\Delta_r Z(k) = \left\{[(Z_1(k) - Z_1(k-1)]^r,\ldots,[Z_p(k) - Z_p(k-1)]^r\right\}.$$

Expanding all of the exponential terms in (3.20) in powers of s_k, multiplying them term by term, and combining terms, yields a more specified form of (3.21); viz.,

$$\ln \Delta\phi(s_k) = \sum_{r=1}^{6}\left[\frac{K_{rk}(is_k)^r}{r!}\right] + o(s_k^6),$$

where

$$K_{1k} = \gamma\mu_k + \lambda_k\theta_k,$$

$$K_{2k} = \sigma^2 + \gamma\sigma_{kk} + \gamma\mu_k^2 + \lambda_k\theta_k^2 + \lambda_k\sigma_k^2,$$

$$K_{3k} = 3\mu_k\sigma_{kk}\gamma + \gamma\mu_k^3 + \lambda_k\theta_k^3 + 3\theta_k\lambda_k\sigma_k^2,$$

$$K_{4k} = 3\gamma\sigma_{kk}^2 + 6\sigma_{kk}\mu_k^2\gamma + \gamma\mu_k^4 + \lambda_k\theta_k^4 + 6\theta_k^2\sigma_k^2\lambda_k + 3\sigma_k^4\lambda_k,$$

$$K_{5k} = 15\sigma_{kk}^2\mu_k\gamma + 10\sigma_{kk}\gamma\mu_k^3 + \gamma\mu_k^5 + \lambda_k\theta_k^5 + 10\lambda_k\theta_k^3\sigma_k^2 + 15\theta_k\lambda_k\sigma_k^4,$$

$$K_{6k} = 15\gamma\sigma^3_{kk} + 45\sigma_{kk}^2\mu_k^2\gamma + 15\mu_k^4\sigma_{kk}\gamma + \mu_k^6\gamma \qquad (4.3)$$

$$+ \lambda_k\theta_k^6 + 15\theta_k^4\lambda_k\sigma_k^2 + 45\theta_k^2\sigma_k^4\lambda_k + 15\sigma_k^6\lambda_k,$$

are the first six cumulants for the kth variable, s_k, $k = 1,2,\ldots,p$; i.e., K_{rk} can be related to the first six moments of the kth component of $\Delta Z(t)$, $r = 1,\ldots,6$. Note that the first two cumulants correspond to the mean and variance of the kth component of $\Delta Z(t)$.

Next define the cumulant vectors

$$K_r = (K_{r1}, K_{r2}, \ldots, K_{rp}),$$

for $r = 3, 4, 5, 6$. Now use the relationships for the cumulants in terms of the sample moments. Let m_{rk} denote the kth component of m_r. Then if $\bar{K}_{rk}$ denotes the sample rth cumulant for the kth component of $\Delta Z(t)$,

$$\bar{K}_{3k} = m_{3k} - 3m_{2k}m_{1k} + 2m_{1k}^3,$$

$$\bar{K}_{4k} = m_{4k} - 4m_{3k}m_{1k} - 3m_{2k}^2 + 12m_{2k}m_{1k}^2 - 6m_{1k}^4,$$

$$\bar{K}_{5k} = m_{5k} - 5m_{4k}m_k - 10m_{3k}m_{2k} + 20m_{3k}m_{1k}^2 + 30m_{2k}^2m_{1k} - 60m_{2k}m_{1k}^3 + 24m_{1k}^5,$$

$$\bar{K}_{6k} = m_{6k} - 6m_{5k}m_{1k} - 15m_{4k} + 30m_{4k}m_{1k}^2 - 10m_{3k}^2 + 120m_{3k}m_{2k}m_{1k}$$

$$- 120m_{3k}m_{1k}^3 + 30m_{2k}^3 - 270m_{2k}^2m_{1k}^2 + 360m_{2k}m_{1k}^4 - 120m_{1k}^6, \qquad (4.4)$$

$k = 1,2,\ldots,p$. Hence form the sample cumulant vectors

$$\bar{K}_r \equiv (\bar{K}_{r1}, \bar{K}_{r2}, \ldots, \bar{K}_{rp}),$$

for $r = 3, 4, 5, 6$. Thus the $\bar{K}_r$ vectors are available from observations. Now matching cumulants setting

$$K_r = \bar{K}_r, \; r = 3, 4, 5, 6,$$

for K_r, and $\bar{K}_r$ defined by (4.3) and (4.4), yields four p-vector equations; i.e., (4p) equations. Combining these with the corresponding ones for the mean vector and covariance matrix, (4.1) and (4.2), provides a total of [5p + p(p + 1)/2] equations (since the covariance matrix has p(p + 1)/2 independent elements). But there are only [4p + 2 + p(p + 1)/2] unknown parameters to be estimated. Hence the system is overdetermined. The problem is easily corrected by matching cumulant vectors only for r = 1, 3, 4, 5 (for r = 2 we use the covariance matrix instead), and then for r = 6 use only the first two elements of the vectors. Thus the complete system of estimation equations may be written in the form:

$$(K_{11},\ldots,K_{1p}) = (\bar{K}_{11},\ldots,\bar{K}_{1p}),$$

$$(K_{31},\ldots,K_{3p}) = (\bar{K}_{31},\ldots,\bar{K}_{3p}),$$

$$(K_{41},\ldots,K_{4p}) = (\bar{K}_{41},\ldots,\bar{K}_{4p}),$$

$$(K_{51},\ldots,K_{5p}) = (\bar{K}_{51},\ldots,\bar{K}_{5p}),$$

$$(K_{61}, K_{62}) = (\bar{K}_{61}, \bar{K}_{62}),$$

$$\text{diag}\left[\sigma^2 + \lambda_1(\theta_1^2 + \sigma_1^2),\ldots,\sigma^2 + \lambda_p(\theta_p^2 + \sigma_p^2)\right] + (\Sigma + \mu'\mu) = S, \quad (4.5)$$

and now there are [4p + 2 + p(p + 1)/2] equations and the same number of unknown parameters to be estimated. The equations are merely algebraic, and may be solved numerically by computer as long as the number of securities in the group, p, does not cause the dimension of the problem to become so large that the capacity of the machine is exceeded.

The estimators obtained by the methods of this section are consistent, but possibly not efficient. However, the sample size, T, can generally be made sufficiently large, at low cost, to produce estimators with small variance.

REFERENCES

[1] Cootner, P. H. 1964. _The Random Character of Stock Market Prices_. Cambridge: The MIT Press.

[2] Doob, J. L. 1953. _Stochastic Processes_. New York: John Wiley and Sons, Inc.

[3] Fama, E. F. 1965. "The Behavior of Stock Market Prices," The Journal of Business, XXXVIII, 1:34-105.

[4] King, B. F. 1966. "Market and Industry Factors in Stock Price Behavior," The Journal of Business, XXXIX, 1, Part II, 139-89.

[5] Mandelbrot, B. 1962. "The Variation of Certain Speculative Prices," IBM Research Note, NC-87. Yorktown Heights, New York: Thomas J. Watson Research Center.

[6] Parzen, E. 1962. Stochastic Processes. San Francisco: Holden-Day, Inc.

[7] Press, S. J. 1967. "A Compound Events Model for Security Prices," The Journal of Business, XL, 3:317-35.

[8] ————. 1968. "A Modified Compound Poisson Process with Normal Compounding," Journal of the American Statistical Association, LXIII, 607-13.

STATISTICAL MODELS
IN DISCRETE DISTRIBUTIONS

J. B. DOUGLAS
University of New South Wales
Australia

SUMMARY

In constructing statistical models of random count phenomena, it is sometimes convenient to start with certain 'elementary' distributions, taken as having known properties, and to combine these systematically to obtain more complex models, the combination being in physically meaningful terms. Such combinations are discussed in this paper; however, there is no discussion of the derivation of distributions directly from stochastic processes.

The ideas involved are illustrated largely by means of a particular example with respect to which interpretations are made of the processes and constants involved.

1. MIXING MODEL

Suppose that during some process a particular kind of impurity is accidentally introduced at a stage which varies according to the steps taken to avoid contamination. The impurity is in the form of a particle of variable size; during the part of the process subsequent to its introduction the particle is (probably) broken into smaller pieces which are then dispersed through the other material, appearing as specks in the finished product. It is these specks which are counted finally, yielding an observed frequency distribution.

To avoid tying the discussion too closely to any particular area, most of the discussion will be put in terms of 'particles' and 'specks.' But these may be 'egg clusters' and the 'larvae' which hatch from them; or 'bombs' and 'bomb fragments'; or 'insurance claims' and their 'amounts'; or 'plants' and their 'leaves'; many applicable examples can be cited without difficulty.

In some circumstances, an 'elementary' distribution may describe reasonably the observed frequency distributions, and its parameters reasonably represent constants of direct interpretation in the physical situation, changes in their values being related in some fashion to changes in the system. In other circumstances the values of such parameters may at best appear to represent some very gross averaging processes not revealing to anyone concerned with the detailed behavior of the system; then, for example, the agreement between the observed and calculated distributions may be systematically poor, as will appear in repeated samples even if they are not large.

In the example described, suppose that, given any fixed size of particle, the number of specks produced from that particle is Poisson-distributed with a mean which, of course, depends on the size of the particle. For mean $u\rho$, where u and ρ will be interpreted in a moment,

$$\Pr(X = x|u) = e^{-u\rho} \frac{(u\rho)^x}{x!}, \quad x = 0,1,2,\ldots,$$

is written for the probability of observing exactly x specks, in other words. But the specks do differ in size.

Suppose for the moment that there are but two sizes, with mean numbers of specks 3ρ and 5ρ, respectively, the relative abundance of the two size distributions being α and $1 - \alpha$, respectively. Rather obviously, the distribution of the overall number of specks observed will be a mixture of the two elementary

Poisson distributions in the proportions α and $1 - \alpha$, with the overall, or unconditional, probabilities of numbers of specks given by

$$\alpha e^{-3\rho} \frac{(3\rho)^x}{x!} + (1 - \alpha)\, e^{-5\rho} \frac{(5\rho)^x}{x!}, \quad x = 0,1,2,\ldots.$$

(In insurance theory the distribution α, $1 - \alpha$ is called the risk function; in a Bayesian context it is the prior distribution.)

More generally, suppose the sizes of the particles vary in any known way (described by an 'elementary' distribution): for the purpose of illustration, take this to be given by a gamma distribution in such a way that u is a realized value from

$$\Pr(U \leq u) = \int_0^u \frac{1}{\Gamma(\kappa)} e^{-u} u^{\kappa-1}\, du, \quad \kappa > 0.$$

(The assumption is arbitrary, like that of Poisson variation already introduced: some facts which show that a gamma variate is not immediately disqualified, at least in some circumstances, as descriptive of size variation are that it is positive, has a single mode, and a long tail to the right.) The mean value of this distribution is κ, the average size of the particles, not in absolute units but meaningful in that a particle of size 2κ produces, on the average, twice as many specks as one of size κ. The unconditional probability distribution of the number of specks produced by a particle considered to be selected at random from this size distribution and then producing specks according to the Poisson assumption is

$$\Pr(X = x) \quad \int_0^\infty e^{-u\rho} \frac{(u\rho)^x}{x!} \frac{1}{\Gamma(\kappa)} e^{-u} u^{\kappa-1}\, du,$$

which after a little algebra becomes

$$\binom{-\kappa}{x} \rho^x (1 + \rho)^{-\kappa - x}, \quad x = 0,1,2,\ldots.$$

This is the familiar (but not 'elementary') <u>negative binomial</u> (or Pascal) distribution, with mean $\kappa\rho$ and variance $\kappa\rho(1 + \rho)$: the process of <u>mixing</u> the Poisson distribution of numbers of specks for fixed particle size—in which the mean and variance are equal—has produced a distribution of numbers of specks in which the variance exceeds the mean. An interpretation of ρ

can now be given: if κ is the average particle size (measured in arbitrary but consistent units) then ρ is the average number of specks overall per average particle.

Hence information about the physical process can be obtained, and control measures, if necessary, can be introduced at the appropriate stage. For instance, a large number of specks may mean a large number of particles is entering the process, or that each particle is generating a large number of specks (perhaps because it enters the process at an early stage); estimates of κ and ρ provide specific numbers relating to these. Even the observation that the overall variance is large compared with the overall mean points to a large number of specks per particle (on this model).

Greater mathematical simplicity is obtained by working with probability-generating functions (PGF's) rather than the probabilities explicitly. If the probabilities are

$$\Pr(X=0) = f_o,\ \Pr(X=1) = f_1,\ldots,\ \Pr(X=x) = f_x,\ldots,$$

the PGF $f(z)$ of X is defined by

$$f(z) = f_o + f_1 z + \ldots + f_x z^x + \ldots = \Sigma z^x f_x$$

$$= \mathcal{E}(z^X).$$

Usually f_x will depend on certain parameters α, $\beta,\ldots$; if $f_x = f_x(\alpha)$ is to be mixed on α, where α has a distribution

$$G(\alpha) = \Pr(A \leq \alpha),$$

the mixed f-distribution with mixing distribution $G(\alpha)$ has probabilities

$$\int f_x(\alpha)\, d\, G\, (\alpha), \quad x = 0,1,2,\ldots,$$

and thus PGF

$$\Sigma\, z^x \int f_x(\alpha)\, d\, G(\alpha) = \int f(z;\alpha)\, d\, G(\alpha),$$

where $f(z;\alpha)$ is the PGF of $f_x(\alpha)$. (The notation

$$\int f(z;\alpha)\, d\, G(\alpha)$$

means either

$$\Sigma\, f(z;\alpha_i)\, \{G(\alpha_i) - G(\alpha_{i-1})\}$$

if α is discrete with jump points $\ldots < \alpha_i < \alpha_{i+1} < \ldots$,

or $\int f(z;\alpha)\, g(\alpha)\, d\alpha$

if α is (absolutely) continuous with density $g(\alpha)$.)
In the preceding example the mixture was given by

$$\int_0^\infty e^{-u\rho + u\rho z} \frac{1}{\Gamma(\kappa)} u^{\kappa-1} e^{-u}\, du$$

$$= (\overline{1+\rho} - \rho z)^{-\kappa} .$$

Of course this account has treated very lightly a number of important points. One is that the particular elementary distributions used have been introduced with almost no justification. What happens if, for example, the 'size' of the particles is supposed Poisson, as would be reasonably postulated if filtering of particles were thought to occur? The mixture in this case is given by the Neyman Type A distribution

$$\sum_{u=0}^{\infty} e^{-u\nu + u\nu z} e^{-\mu} \frac{\mu^u}{u!} = \exp(-\mu + \mu e^{-\nu + \nu z}),$$

where the notation has been altered to avoid confusion: ν (instead of ρ) is the average number of specks per average particle, and μ the mean number of particles entering the process at a particular point. (Notice that the particle 'size' now proceeds in integral multiples of ν: $0, \nu, 2\nu, 3\nu, \ldots$.) Once more the overall mean number of specks is $\mu\nu$ and the variance, $\mu\nu\,(1 + \nu)$, but there is little other detailed correspondence. For example, in the overall distribution of specks, the third factorial cumulant is $2\kappa\rho^3$ for the gamma mixing distribution and $\mu\nu^3$ for the Poisson; similarly, the fourth factorial cumulants are $6\kappa\rho^4$ and $\mu\nu^4$, respectively. The PGF's are visibly different, and so are the probabilities: those of the negative binomial cannot have more than a single mode, while those of the Neyman Type A may have any number, for example.

Of mixed distributions, much attention has been given to the class known as the <u>mixed Poisson</u>: a number of general properties can be established for this class which throw light on the kinds of structure involved.

For this class, the PGF has the form

$$\int e^{-u\nu + u\nu z}\, d\,G(u)$$

(all three examples already quoted are special cases), which can be written as

$$\int (e^{-\nu} + \nu z)^u \, d\, G(u).$$

Hence, introducing the PGF of G(u) as

$$g(z) = \mathcal{E}(z^U) = \int z^u \, d\, G(u),$$

the PGF of the mixture can be written as

$$g(e^{-\nu} + \nu z) = h(z), \quad \text{say},$$

where—although they are not shown explicitly—the function $g(\cdot)$ will usually carry additional parameters with it (α, ζ and μ in the previous examples).

It is true in general that the variance of the mixture exceeds its mean (strictly, except in a degenerate case), and thus a mixed Poisson can never be a Poisson distribution (for which the mean and variance are equal). The mean and variance of the mixed variate X can be written down from the relations

$$\mathcal{E}(X) = h'(1),$$

$$V(X) = h''(1) + h'(1) - \{h'(1)\}^2,$$

which are valid for any variate; they lead to

$$\mathcal{E}(X) = \nu\, \mathcal{E}(U)$$

and $$V(X) = \nu^2\, V(U) + \mathcal{E}(X)$$

$$> \mathcal{E}(X)$$

(unless $V(U) = 0$, in which case the equality obtains, but the "mixing" distribution is then degenerate with one jump point at which all its probability is located). It also follows that if $\nu > 1$

$$\mathcal{E}(X) > \mathcal{E}(U) \quad (> 0)$$

and $$V(X) > V(U) .$$

It is clearly of interest whether different mixing distributions can give rise to the same mixture; if they can, disentangling the structure from observations on the mixture alone will be impossible. But if $g_1(z)$ and $g_2(z)$ are the PGF's of different mixing distributions, those of their mixed Poisson distributions are

$$g_1(e^{-\nu + \nu z}) \text{ and } g_2(e^{-\nu + \nu z}),$$

and these can be the same only if $g_1 = g_2$. (Unfortunately, from this point of view, it is not the case that a corresponding result holds for general mixtures, nor necessarily for other methods of combining distributions; but of course a model which admitted only one interpretation might also be of correspondingly limited applicability.) Indeed, much more than this can be asserted. Supposing the PGF of the mixture to be $g(e^{-\nu + \nu z})$, its factorial moment-generating function (MGF) is $g(e^{\nu t})$, obtained by replacing z by 1 + t. But the MGF of the mixing distribution is $g(e^{t'})$, by replacing z by $e^{t'}$ in its PGF, and hence, writing $t' = \nu t$, the rth factorial moment of the mixture = ν^r times the rth power moment of the mixing distribution.
(This is the generalization of the mean and variance relations obtained earlier, and illustrates the appropriateness of factorial, rather than power, functions for discrete variates.) Thus, given a scale factor ν (which may be obtainable from some information regarding the mean of the mixing distribution), higher moments of the mixing distribution be estimated from observations on the mixture. Even a uniqueness theorem holds (Maeeda [8]): a variate is a mixture of Poisson variates if and only if its factorial MGF is the MGF of some non-negative variate. (Quite generally, without identifiability, no such estimation is possible; see Choi and Bulgren [4]) for some convenient references, especially to Teicher, on this).

Among the most useful properties of a family of distributions is a simple convolution relation: if X_1 and X_2 are independent variates belonging to the family, it is often physically sensible to expect $X_1 + X_2$, perhaps subject to some restrictions, also to be in the family. Thus, if X_i are independent normal variates with means μ_i and variances σ_i^2, i = 1,2 (written $IN(\mu_i, \sigma_i^2)$), it is well known that $X_1 + X_2$ is $N(\mu_1 + \mu_2, \sigma_1^2 + \sigma_2^2)$; if X_i are independently binomial with means $\nu_i\tilde{\omega}$, $X_1 + X_2$ is also binomial with mean $(\nu_1 + \nu_2)\tilde{\omega}$: notice that $\tilde{\omega}$, the probability of a "success" in the conventional labeling, must be the same for both variates. In terms of PGF's, if X_1 and X_2 are independent variates with PGF's $h_1(z)$ and $h_2(z)$, respectively, that of $X_1 + X_2$ is

$$\mathcal{E}(z^{X_1 + X_2}) = \mathcal{E}(z^{X_1})\ \mathcal{E}(z^{X_2}) = h_1(z)\ h_2(z),$$

the product of the PGF's. In the present case, if $g_i(z)$, $i = 1,2$, are the PGF's of the mixing distributions on the (same) Poisson family $e^{-u\nu + u\nu z}$, the PGF of the convolution of the mixing distributions is $g_1(z)\ g_2(z)$; replacing z by $e^{-\nu + \nu z}$ gives the result:

> the convolution of two independent mixed Poisson variates mixed on the same Poisson family is itself a mixed Poisson variate with mixing distribution the convolution of the two mixing distributions.

So, for example, if particles were entering the system already described with two independent size distributions ($g_1(z)$ and $g_2(z)$), where the number of specks produced per particle of the same size was the same (given by $e^{-u\nu + u\nu z}$), the total number of specks would still be given by a mixed Poisson distribution. (This "reproductive" property belongs to general mixtures, by the same kind of argument.)

To be still more specific, suppose that the particles entering have a size distribution (going up in integral multiples) given by a Poisson distribution with mean μ, and that the number of specks per particle of size $u\nu$, $u = 0,1,2...$, is also Poisson. The number of specks overall has already been shown to have, in this case, a Neyman Type A distribution with PGF:

$$e^{-\mu + \mu e^{-\nu + \nu z}};$$

what has just been said is that if, as well, there are two size distributions for particles both Poisson but with means μ and μ' and mean μ' applies, then the number of specks overall will have the PGF

$$e^{-\overline{\mu + \mu'} + \overline{\mu + \mu'}\, e^{-\nu + \nu z}},$$

still a Neyman Type A distribution with parameters $\mu + \mu'$, ν in place of μ,ν. (The Poisson assumption for the two size distributions is of course not necessary for the convolution result to hold—only for the neat reproductive result.)

Alternatively, going beyond the convolution result stated, consider the case where the size distribution of the entering particles is the same for both kinds of particles (the probabilities associated with $u = 0,1,2,...$ are given by a Poisson distribution with mean μ), but the numbers of specks per particle of given size differ for the two kinds, being Poisson-distributed with means ν and ν', respectively. The overall distribution of the number of specks $X_1 + X_2$ thus has the PGF

$$\exp(-\mu + \mu e^{-\nu + \nu z}) \exp(-\mu + \mu e^{-\nu' + \nu' z})$$

$$= \exp\{-2\mu + \mu(e^{-\nu + \nu z} + e^{-\nu' + \nu' z})\},$$

and this is a three parameter Neyman Type A distribution. In practice, distributions with more than two parameters have not been used very widely; perhaps this is because of uncertainty associated with the interpretation of their parameters rather than because of the algebraic complexity of the distributions: in the case of the Poisson-Binomial distribution with PGF

$$\exp\{-\mu + \mu(\chi + \tilde{\omega} z)^{\nu}\}$$

for example, an interpretation of ν in a physical context is certainly rare.

2. RANDOM STOPPING MODEL

Another physically meaningful process considers circumstances where each particle (numbered 1,2,3,... for reference) which enters the system produces specks $X_1, X_2, X_3, \ldots$ with the same probability function (PGF $\phi(z)$), but the number N of particles entering has itself a PGF given by

$$\gamma(z) = \Sigma z^n \Pr(N=n).$$

Then the total number of specks is

$$X_1 + X_2 + \ldots + X_n,$$

N being a random variable, and by precisely the mixture argument as before, the PGF of this sum, randomly stopped according to the PGF $\gamma(z)$, is

$$\Sigma \text{ (PGF of X for fixed n) } \Pr(N=n)$$

$$= \Sigma\{\phi(z)\}^n \Pr(N=n)$$

$$= \gamma\ (\phi(z)).$$

Thus a stopped distribution is a special kind of mixed distribution, although physically—where the interpretation of a PGF is not usually direct—the processes are not conveniently thought of in this way.

Because of the special form of this kind of mixture, it deserves separate study, and a first illustration can be given: obviously

$$\gamma(\phi(z;\nu)) \text{ is } g(e^{-\nu + \nu z})$$

for $\phi(z;\nu) = e^{-\nu + \nu z}$ with $\gamma = g$.

Hence, a mixed Poisson distribution with weight function given by g(z) is also a g-stopped Poisson distribution (of course special forms for g may also produce similar identities). A very special case is the Neyman Type A distribution, where $g(z) = e^{-\mu + \mu z}$ also corresponds to a Poisson distribution: thus it can be generated as a mixture as

$$\sum_{u=0}^{\infty} (e^{-\nu + \nu z})^u \, e^{-\mu} \frac{\mu^u}{u!}$$

or as a stopped distribution as that of $X_1 + X_2 + \ldots + X_N$, where X_i: Poisson (ν) and N: Poisson (μ).

Another specific model considers a system in which, for given size $u\nu$ of particle, the numbers entering the system are Poisson $e^{-u\nu + u\nu z}$, and the scaled size U is supposed uniformly distributed on (0,1). Thus the numbers entering the system averaged over the size distribution have the PGF (see Anscombe [1])

$$\int_0^1 e^{-u\nu + u\nu z} \, du$$

$$= \frac{e^{-\nu + \nu z} - 1}{-\nu + \nu z} :$$

if then the number of specks per particle is Poisson (μ), the distribution of the overall number of specks is

Poisson (μ)-stopped (mixed Poisson ($u\nu$) by the unit uniform) with PGF

$$\exp\left\{-\mu + \mu \frac{e^{-\nu + \nu z} - 1}{-\nu + \nu z}\right\} ,$$

and this is a two-parameter Neyman Type B distribution. (It is interesting to notice that the use of a uniform distribution on $(0,\alpha)$ produces no change: the form is then

$$\exp \{-\mu + \mu \frac{e^{-\alpha\nu + \alpha\nu z} - 1}{-\alpha\nu + \alpha\nu z}\},$$

with neither α nor ν separately estimable from the final frequency distribution alone.) Alternatively, a Poisson (μ)-stopped uniform variate on (0,1) has the PGF

$$\exp (-\mu + \mu \frac{z - 1}{\ln z})$$

corresponding to particles formed by the aggregation of Poisson (μ) distributed numbers of 'sub-particles' from a uniform (0,1) size distribution; if this is formally applied as the stopping distribution to a Poisson (ν) variate this gives the

(Poisson (μ)-stopped unit uniform)—stopped Poisson (ν),

with PGF again the Neyman Type B.

Again the fundamental character of the Poisson distribution appears: Poisson-stopped distributions (or sometimes "stuttering Poisson" distributions) are as important as mixed Poisson distributions. (Notice that although "stopping" is a special kind of "mixing," a Poisson-stopped distribution is not usually a special case of the mixed Poisson distribution: in the Poisson-stopped case the Poisson is the mixing, not the mixed, distribution. And the infinitely divisible property of the Poisson-stopped distribution is of great theoretical importance.) The general form of the Poisson-stopped PGF is

$$H(z) = e^{-\mu + \mu \phi(z)},$$

where $\phi(z)$ is the PGF of the component variate Y.

Along the previous lines, writing Y* for the Poisson-stopped f-variate,

$$\mathcal{E}(Y^*) = \mu \, \mathcal{E}(Y)$$

and $$V(Y^*) = \mathcal{E}(Y^*) \frac{\mathcal{E}(Y^2)}{\mathcal{E}(Y)}.$$

The variance of a Poisson-stopped variate thus may or may not exceed its mean: of course it does if Y is itself Poisson, since it is then also a mixed Poisson variate. Since it is often true that $\mathcal{E}(Y^2) > \mathcal{E}(Y)$ (always for variates on 0,1,2,... for which $\Pr(Y>1) > 0$), it is also often true that $V(Y^*) > (Y^*)$. Writing

$$\phi(z) = \sum_{0} z^y \phi_y,$$

obviously,

$$\sum_{2} (y-1)\,\phi_y > 0,$$

and so

$$-(1-\sum_{o} \phi_y) + \sum_{1} (y-1)\,\phi_y > 0;$$

i.e., $-1 + \phi_o > -\sum_{o} y\phi_y$.

Multiplying by μ and exponentiating,

$$e^{-\mu + \mu \Pr(Y=0)} > e^{-\mu\,\mathcal{E}(Y)},$$

where the right-hand side is the probability of the zero class for a Poisson variate with the same mean as that of the stopped variate: there is more in the case of the stopped variate.

The natural convolution property holds here too: suppose in $X_1 + X_2 + \ldots + X_{N_1}$ each X has the PGF $\phi(z)$ and N_1 is Poisson (μ_1), and in $Y_1 + Y_2 + \ldots + Y_{N_2}$ each Y has the PGF $\phi(z)$ and N_2 is Poisson (μ_2). Then the distribution of

$$X_1 + \ldots + X_{N_1} + Y_1 + \ldots + Y_{N_2}$$

is given by

$$e^{-\overline{\mu_1 + \mu_2} + \overline{\mu_1 + \mu_2}\,\phi(z)} :$$

note that the same $\phi(z)$ appears in both places, corresponding to the identity of distribution of the numbers of specks per particle for both kinds of particles even though the rate of occurrence of particles may be different.

Following these remarks, a more abbreviated description of modeling processes for the problem introduced earlier is given, the steps being set out in succession, being terminated at a fourth stage quite arbitrarily. Of the many formulations possible, only a small number is exhibited (together with some specific examples): these are illustrative and not comprehensive. (A longer, but still quite incomplete, catalogue is to be found in Rao, Mitra and Matthai [12].)

To save repetition in the table,

$X|\alpha$ means the random variable X at that stage, given the value α of some variable;

X: $g_\alpha(z)$ means X has the PGF $g_\alpha(z)$;

and X is the number of specks, the observable random variable, at each stage, the previous distributional assumptions being carried down in successive column entries.

For completeness, the alternative common derivation of the negative binomial will be given in this way :

1. Specks/particle of size u logarithmic ($u\nu$)

$$X: \frac{\ln(1 - u\nu z)}{\ln(1 - u\nu)} .$$

2. n such particles

$$X: \frac{\ln(1 - u\nu z)}{\ln(1 - u\nu)}^{n} .$$

(First kind of Stirling distribution)

3. N such particles, $N: e^{-\lambda+\lambda z}$

$$X: \left(\frac{1}{1-u\nu} - \frac{u\nu}{1-u\nu} z\right)^{\lambda/\ln(1 - u\nu)}.$$

(Negative binomial)
(Poisson stopped logarithmic)
(Mixed First Kind of Stirling by Poisson)

The "stopping formula" $\gamma(\phi(z))$ for a γ-stopped ϕ-distribution yields a PGF even if the γ-variate is not integral-valued, although of course the interpretation that it is the PGF of

$$X_1 + X_2 + \ldots + X_N,$$

where $X_1, X_2, \ldots$ are identically distributed independent variates with PGF $\phi(z)$ and N the PGF of $N(= 0,1,2,\ldots)$, is not then valid. Thus, e.g., if

$\phi(z) = e^{-\mu + \mu z}$, corresponding to a Poisson variate, and

$\gamma(z) = (1 - \ln z)^{-\beta}$, corresponding to a gamma variate,

then

$$\gamma(\phi(z)) = (\overline{1 + \mu} - \mu z)^{-\beta}$$

corresponds to a negative binomial variate, but though it may be so <u>described</u>, it cannot be <u>interpreted</u> as a gamma-stopped Poisson variate: $(1 - \ln z)^{\beta}$ does not have a power series expansion in z, for it does not correspond to an integral-valued variate. (See Molenaar [9] for an extension to characteristic functions.) The general interpretation as a special mixture remains:

1. Specks/particle of size u Poisson ($u\nu$)

X: $e^{-u\nu + u\nu z}$

2(a) 1 particle, size distribution Poisson (μ)

U: $e^{-\mu + \mu z}$

X: $\Sigma\, e^{-u\nu + u\nu z}\, e^{-\mu} \frac{\mu^x}{x!}$

$= e^{-\mu + \mu e^{-\nu + \nu z}}$

(Neyman Type A)
(Mixed Poisson by Poisson)

3(a) n such particles

X: $e^{-n\mu + n\mu e^{-\nu + \nu z}}$

(Neyman Type A)

4(a) N such particles, N Poisson (λ)

N: $e^{-\lambda + \lambda z}$.

X: $e^{-\lambda + \lambda e^{-\mu + \mu e^{-\nu + \nu z}}}$.

(Poisson-stopped Neyman Type A)

2(a') 1 particle, size distribution gamma

U: $(1 - \ln z)^{-\mu}$

X: $\int_0^\infty e^{-u\nu + u\nu z} \frac{1}{\Gamma(\mu)} e^{-u}\, u^{\mu-1} du$

$= (\overline{1 + \nu} - \nu z)^{-\mu}$
Negative binomial)
(Mixed Poisson by gamma)

3(a') n such particles

X: $(\overline{1 + \nu} - \nu z)^{-n\mu}$

(Negative binomial)

4(a') N such particles, N: Poisson (λ)

N: $e^{-\lambda + \lambda z}$

X: $\exp(-\lambda + \lambda(\overline{1 + \nu} - \nu z)^{-\mu})$.

(Poisson - Pascal)

2(b) n such particles

X: $e^{-nu\nu + nu\nu z}$

(Poisson)

3(b) N such particles, N Poisson (λ)

N: $e^{-\lambda + \lambda z}$

X: $e^{-\lambda + \lambda e^{-u\nu + u\nu z}}$

(Neyman Type A)
(Poisson-stopped Poisson)

3(b') N such particles, N Neyman Type A

N: $e^{-\lambda + \lambda^{-\mu + \mu z}}$

X: $e^{-\lambda + \lambda e^{-\mu + \mu e^{-u\nu + u\nu z}}}$

(Neyman Type A-stopped Poisson)

4(b) N such particles (all of size u)

U: $e^{-\mu + \mu z}$

X: $\Sigma\, e^{-\lambda + \lambda e^{-u\nu + u\nu z}}\, e^{-\mu} \frac{\mu^u}{u!}$.

(Mixed Neyman Type A (λ,ν) on ν by Poisson)

1. A single particle of size u enters a system.
The particle produces $X|u$ specks, with

$$X|u: \quad f_\nu\ (z|u).$$

2(a) 1 particle enters the system with size u given by

$Pr(U \leq u) = G_\mu(u)$, $U: \ g_\mu(z)$.

The unconditional number of specks

$X: \ \int f_\nu(z|u)\ d\ G_\mu(u)$.

mixed-f by g

3(a) n such particles independently enter the system.

Hence $X_1 + X_2 + \ldots + X_n = X$:

$$\{\int f_\nu\ (z|u)\ d\ G_\mu(u)\}^n$$

n-convolution of (mixed-f by g)

4(a) N such particles enter the system where

$Pr(N \leq n) = H_\lambda\ (n)$, $N: \ h_\lambda(z)$.

Hence

$X_1 + X_2 + \ldots + X_N = X$:

$$\int \{\int f_\nu(z|u)\ d\ G_\mu(u)\}^n d\ H_\lambda(n)$$

$$= h_\lambda\ \{\int f_\nu(z|u)\ d\ G_\mu(u)\}$$

h-stopped (mixed-f by g)

2(b) n particles of size u enter the system.

They independently produce specks

$X_1, X_2, \ldots, X_n$.

Hence

$X_1 + X_2 + \ldots + X_n = X: \quad \{f_\nu(z|u)\}^n$.

n-convolution of f

3(b) N such particles enter, where

$Pr(N \leq n) = H\lambda(n)$, $N: \ h_\lambda\ (z)$.

Hence $X_1 + X_2 + \ldots + X_N = X$:

$$\int \{f_\nu(z|u)\}^n\ d\ H_\lambda(n) = h_\lambda(f_\nu(z|u)\ .$$

h-stopped f

4(b) To introduce a size distribution on single particles, return must be made to 1. Alternatively, if all N particles have the same size u, averaging over size of particle generating X specks (as in 3(b)) gives

$$\int h_\lambda \{f_\nu\ (z|u)\}\ d\ G_\mu(u).$$

(h-stopped f) mixed by g

$$\int_0^\infty \{\phi(z)\}^u \frac{1}{\Gamma(\beta)} e^{-u} u^{\beta-1} du$$

$$= \gamma(\phi(z)),$$

i.e., a mixed Poisson with gamma weights. The alternative interpretation, given earlier, of course is still valid: $\gamma(\phi(z))$ can be obtained as a Poisson-stopped logarithmic distribution, where in the above notation the Poisson PGF is

$$\exp\{-\beta \ln (1 + \mu) + \beta z \ln (1 + \mu)\} ,$$

and the logarithmic

$$\frac{\ln (1 - \frac{\mu}{1 + \mu} z)}{\ln (1 - \frac{\mu}{1 + \mu})} .$$

This illustrates that the parametric specification depends on the point of view, the specification through β and μ not appearing likely to be useful if the components are of primary interest.

3. FURTHER COMBINATIONS: COMPARISONS OF MODELS

As in the schematic representation, repetitions of the combining processes may be carried out, although the multiplication of parameters makes their use quickly less plausible unless a specific application is in mind.

A very natural structure, which could be called the _triple Poisson_ by analogy with the _double Poisson_ for the Neyman Type A distribution, obtainable either by mixing or stopping or a combination of these using Poisson distributions at each stage, has the PGF

$$\exp[-\lambda + \lambda \exp\{-\mu + \mu \exp (-\nu + \nu z)\}],$$

where λ, μ, ν are positive. An illustration of its genesis can be given with reference to insects laying egg-clusters over a nonhomogeneous field (for particles and specks the translation is obvious). Suppose a field is divided into homogeneous areas within which the mean number of egg clusters laid per quadrat is one of

$$0, \mu, 2\mu, \ldots, i\mu, \ldots,$$

(where for given $i\mu$ the variation is Poisson) and areas selected

$$\kappa_1 = \lambda\mu\nu$$

$$\kappa_2 = \lambda\mu\nu\ (1 + \nu + \mu\nu)$$

$$\kappa_3 = \lambda\mu\nu\ (1 + 3\nu + 3\mu\nu + \nu^2 + 3\mu\nu^2 + \mu^2\nu^2)$$

$$\kappa_4 = \lambda\mu\nu\ (1 + 7\nu + 7\mu\nu + 6\nu^2 + 18\mu\nu^2 + \nu^3 + 7\mu\nu^3 + 6\mu^2\nu^2 + 6\mu^2\nu^3 + \mu^3\nu^3),$$

from $\kappa_r = \lambda \sum_{n=1}^{r} \mathscr{S}_r^{(m)} \nu^m \sum_{n=1}^{m} \mathscr{S}_m^{(n)} \mu^n$.

Introducing $\alpha = \lambda e^{-\mu + \beta}$, $\beta = \mu e^{-\nu}$, explicit formulae for the first five P_x^* are:

$$P_o^* = e^{\alpha-\lambda}$$

$$P_1^* = e^{\alpha-\lambda}\ \alpha\beta\nu$$

$$P_2^* = e^{\alpha-\lambda}\ \alpha\beta \frac{\nu^2}{2!} \{\alpha\beta + 1 + \beta\}$$

$$P_3^* = e^{\alpha-\lambda}\ \alpha\beta \frac{\nu^3}{3!} \{\alpha^2\beta^2 + 3\alpha\beta\ (1 + \beta) + 1 + 3\beta + \beta^2\}$$

$$P_4^* = e^{\alpha-\lambda}\ \alpha\beta \frac{\nu^4}{4!} \{\alpha^3\beta^3 + 6\alpha^2\beta^2\ (1 + \beta) + \alpha\beta\ (7 + 18\beta + 7\beta^2) + 1 + 7\beta + 6\beta^2 + \beta^3\},$$

but it is not advantageous to work directly with these expressions. A recurrence relation derived from the representation as a Poisson-stopped Neyman Type A distribution is relatively simple:

$$P_{x+1}^* = \frac{\lambda}{x+1} \sum_{p=0}^{x} p\ P_p\ P_{x+1-p}^*,$$

where P_p is the Neyman Type A probability given by

$$P_p = e^{-\mu+\beta} \frac{\nu^p}{p!} \sum_{q=1}^{p} \mathscr{S}_p^{(q)} \beta^q,$$

with $P_o = e^{-\mu + \beta}$.

Using the sample factorial cumulants, the same functions of the sample moments as the factorial cumulants are of distribution so that the probability of selecting i is given by a Poisson distribution with mean λ. Then the number of clusters per quadrat follows a Neyman Type A (λ, μ) distribution; if the number X of eggs per cluster is now supposed Poisson (ν) the stopped distribution of

$$X_1 + X_2 + \ldots + X_N,$$

where N has the Neyman Type A distribution, is that described.

An examination of some of its properties helps illuminate the problems of distinguishing between models. Its factorial cumulants are readily given, since the factorial cumulant-generating function is

$$- \lambda + \lambda e^{-\mu + \mu e^{\nu t}}$$

$$= -\lambda + \lambda e^{\mu(e^{\nu t}-1)}$$

$$= -\lambda + \lambda \, \Sigma \, \mu^n \frac{(e^{\nu t}-1)^n}{n!} ,$$

and use of the expansion for Stirling Numbers $\mathcal{S}_m^{(n)}$ of the Second Kind (Jordan [7]),

$$\frac{(e^t - 1)^n}{n!} = \sum_{m=n}^{\infty} \mathcal{S}_m^{(n)} \frac{t^m}{m!} ,$$

leads to

$$\lambda \sum_{m=1}^{\infty} \nu^m \frac{t^m}{m!} \sum_{n=1}^{m} \mathcal{S}_m^{(n)} \mu^n$$

and hence to the factorial cumulant $\kappa_{(m)}$,

$$\kappa_{(m)} = \lambda \, \nu^m \, \Sigma \, \mathcal{S}_m^{(n)} \mu^n .$$

Thus $\kappa_{(1)} = \lambda\mu\nu$

$$\kappa_{(2)} = \lambda\mu\nu^2 (1 + \mu)$$

$$\kappa_{(3)} = \lambda\mu\nu^3 (1 + 3\mu + \mu^2)$$

$$\kappa_{(4)} = \lambda\mu\nu^4 (1 + 7\mu + 6\mu^2 + \mu^3)$$

or

moments, moment-like estimators λ_m, μ_m, ν_m are given by

$$K_{(1)} = \lambda_m \mu_m \nu_m,$$

$$K_{(2)} = \lambda_m \mu_m \nu_m^2 (1 + \mu_m),$$

$$K_{(3)} = \lambda_m \mu_m \nu_m^3 (1 + 3\mu_m + \mu_m^2).$$

The solution of these is:

(μ): μ_m and $\frac{1}{\mu_m}$ as positive roots of

$$\mu_m^2 - \frac{3 - 2L}{L - 1} \mu_m + 1 = 0, \quad 1 < L = \frac{K_{(1)}K_{(3)}}{K_{(2)}^2} < \frac{5}{4},$$

$$(\nu): \quad \nu_m = \frac{K_{(2)}}{K_{(1)}(1 + \mu_m)} \text{ and } \frac{K_{(2)}}{K_{(1)}(1 + 1/\mu_m)},$$

and

$$(\lambda): \quad \lambda_m = \frac{K_{(1)}^2}{K_{(2)}} (1 + 1/\mu_m) \text{ and } \frac{K_{(1)}^2}{K_{(2)}} (1 + \mu_m),$$

where all three of the first or second members of the pairs must be taken. It is not clear how to choose the 'correct' triplet without reference to the model and some information from intermediate stages of its construction: indeed, as will be seen immediately, rather different triplets may give rather similar probability distributions. (Maximum likelihood equations can be obtained through the recurrence relation quoted; but it is much harder to try to see the behavior of these estimators.) Certain limiting results may throw some light on the behavior, and the simplest for small λ, μ and ν are as follows. Since

$$e^{-\lambda + \lambda e^{-\mu + \mu e^{-\nu + \nu z}}} = e^{-\lambda\mu\nu + \lambda\mu\nu z} + O(\lambda\mu\nu^2),$$

as $\nu \to 0$ with $\lambda\mu\nu$ finite, the distribution is approximated by the Poisson ($\lambda\mu\nu$) form. If $\lambda\mu$ is finite as $\mu \to 0$, the approximation is that of a Neyman Type A ($\lambda\mu,\nu$) form; if $\lambda \to 0$ the approximation is that of a Neyman Type A ($\lambda\mu,\nu$) form; if $\lambda \to 0$ the approximation is given by

$$(1 - \lambda) + \lambda e^{-\mu + \mu e^{-\nu + \nu z}},$$

that of a Neyman Type A distribution with additional zero class probability. (It might be repeated that the various models of this distribution, including, for example,

Poisson (λ)-stopped Neyman Type A (μ,ν),
Neyman Type A (λ,μ)-stopped Poisson (ν),
mixed Neyman Type A ($u\mu,\nu$) by Poisson (λ) on u,

are all equally valid: what model is used—and hence the interpretation of the parameters even if the estimation procedure for the parameters in the final distribution is the same—will depend on the structure of the process it is intended to model.)

There are various levels at which one wishes to distinguish between models. Perhaps the most common, and the most superficial, is simply in order to reproduce reasonably closely sets of observed frequencies with different parametric probability functions, hopefully classifying in some way the type of data fitted for which there is goodness of agreement. No interpretation of the probability model is used other than by comparing long run frequencies and probabilities, and no use is made of the structure of the model. Goodness-of-fit tests such as chi-square are typically used here; and it is possible, for example, to evaluate the power of such a test, e.g., when the hypothesis tested is a Neyman Type A and the alternative a negative binomial distribution. Various criteria, such as the variance-to-mean ratio, or the ratios of successive factorial cumulants, can also be used; however, none of these are being considered here.

Obviously, even supposing that at the above level a Neyman Type A distribution were accepted as appropriate, no information about the structure is thus conveyed. As the simplest structures, the competing models in this case are:

A Poisson-distributed random number of identical particles, each producing a Poisson-distributed number of specks:

X_i: Poisson (ν), $i = 1,2,\ldots, N$,

N: Poisson (μ),

whence $X_1 + X_2 + \ldots + X_N$: $e^{-\mu + \mu e^{-\nu + \nu z}}$.

A Poisson-distributed random number of specks per particle itself with Poisson-distributed size:

$$X|u: \quad e^{-u\nu + u\nu z},$$

$$U: \quad e^{-\mu + \mu z},$$

$$\text{whence } X: \quad e^{-\mu + \mu e^{-\nu + \nu z}}.$$

In the former case, in each of n repetitions of the 'experiment' which leads to the frequency distribution with n observations, many particles are supposed to be producing specks the total number (or some constant proportion) of which is observed: the average number of such particles is μ and for each particle the average number of specks is ν. If this is meaningful, it may be possible to estimate in some direct fashion the mean number of particles and/or the mean number of specks per particle; even without detailed individual observations on the separate X_i and N, some assessment of the adequacy of the Poisson assumptions may be able to be made.

In the second case, in each of the n repetitions, one particle is supposed to be chosen from a size distribution, the observation being of the number of specks this particle produces. Here too estimation on the size distribution directly and/or on the number of specks per particle will provide similar information. Most to the point, however, is the analysis of the experimental circumstances so that one knows what one is looking for. If there is a size distribution which cannot reasonably be replaced by the mean size, the former model is simply not relevant, for instance. (Perhaps the first application of these ideas was in Newbold [10], relative to accident data: see also Arbous and Kerrich [2].)

Even more significant from an operational view may be a consideration of the effects of changes: absolute values may well be unknown, but relative values known. In the former case, doubling the mean number of particles producing specks gives the PGF of the new number of specks as

$$e^{-2\mu + 2\mu e^{-\nu + \nu z}}:$$

fresh estimation should approximately double the estimate for μ, leaving that for ν unchanged; on the other hand, doubling the speck-producing capacity of each particle (doubling the mass, or surface area, or diameter?) should give

$$e^{-\mu + \mu e^{-2\nu + 2\nu z}}.$$

Similar considerations naturally apply to this latter case. (See Pielou [11].)

Another aspect of the difficulties of proceeding directly and solely from the single observed distribution may be illustrated by the 'triple Poisson' already quoted. Here, for example, the distributions specified by

$$(\lambda,\mu,\nu) = (\tfrac{3}{2}, \tfrac{1}{2}, \tfrac{4}{3})\ (\tfrac{3}{4}, 2, \tfrac{2}{3}),\ (\tfrac{20}{39}, 39, \tfrac{1}{20})$$

all have means of 1 and variances of 3 (factorial second cumulants 2); the first two also have as third factorial cumulant $4\frac{8}{9}$ and the third $4\frac{199}{400}$. Given a set of observations with sample factorial first, second and third cumulants near 1, 2 and $4\frac{1}{2}$, respectively, what hope could one have of estimating accurately λ, μ, ν from samples of reasonable size? Especially with the high variability of third moments in mind, it is unrealistic to expect equivalents of the method of moments to be successful; any other method would need separate investigation, but the following tabulation of the probability functions (carried out by W. M. Stiteler of the Department of Statistics, The Pennsylvania State University) makes it very implausible that <u>any</u> method will be very effective. Because some 60% of the distributions are in the zero class (chosen so that the distributions would be short) there is less opportunity for differences to show up in the later parts of the distributions; but graphs are given both on ordinary and semilogarithmic graph paper to exhibit these differences as prominently as possible. Given the considerable differences between the first two triplets (λ,μ,ν) and the exceedingly large differences between these and the third, the near correspondence of the probabilities is striking.

Another example which illustrates the impossibility of making inferences regarding the structure merely from the observed frequency distribution does not go beyond the Poisson and binomial distributions. If the binomial

$$(\chi + \tilde{\omega}z)^{\nu}$$

is mixed on ν by a Poisson (μ), the mixed binomial distribution is

$$\Sigma(\chi + \tilde{\omega}z)^{\nu}\, e^{-\nu}\, \frac{\mu^{\nu}}{\nu!}$$

$$= e^{-\mu\tilde{\omega} + \mu\tilde{\omega}z},$$

a Poisson ($\mu\tilde{\omega}$) distribution. I.e., if observations are made on a binomial variate whose exponent ν varies from experiment to experiment according to a Poisson distribution, the number of "successes" will be Poisson-distributed with parameter the product of the primary Poisson parameter and the probability of a single "success."

"Triple Poisson" Probabilities

$$e^{-\lambda + \lambda e^{-\mu + \mu e^{-\nu + \nu z}}}$$

	(λ,μ,ν)		
	$(\frac{3}{2}, \frac{1}{2}, \frac{4}{3})$	$(\frac{3}{4}, 2, \frac{2}{3})$	$(\frac{20}{39}, 39, \frac{1}{20})$
100 Pr(X=0)	63.00	62.71	64.64
1	11.49	12.17	9.18
2	9.72	9.40	9.39
3	6.46	6.30	6.96
4	3.92	3.94	4.34
5	2.32	2.36	2.49
6	1.35	1.37	1.38
7	0.77	0.78	0.76
8	0.44	0.44	0.41
9	0.24	0.24	0.22

Neither of these two parameters graphically demonstrated on the next two pages is separately estimable; and in any of the previous arguments where a Poisson distribution appeared it would of course be replaced by an appropriate mixed-binomial by Poisson distribution, to the complication of the model. (An equivalent statement concerning the distribution is that it is a Poisson-stopped binomial ($\chi + \tilde{\omega}z$) distribution.) Nor is this behavior peculiar to the Poisson distribution: for the geometric distribution with PGF

$$g(z) = \frac{1-\alpha}{1-\alpha z} = (1 - \alpha)\ \Sigma\ \alpha^x z^x,$$

the mixture (or stopped distribution) has PGF

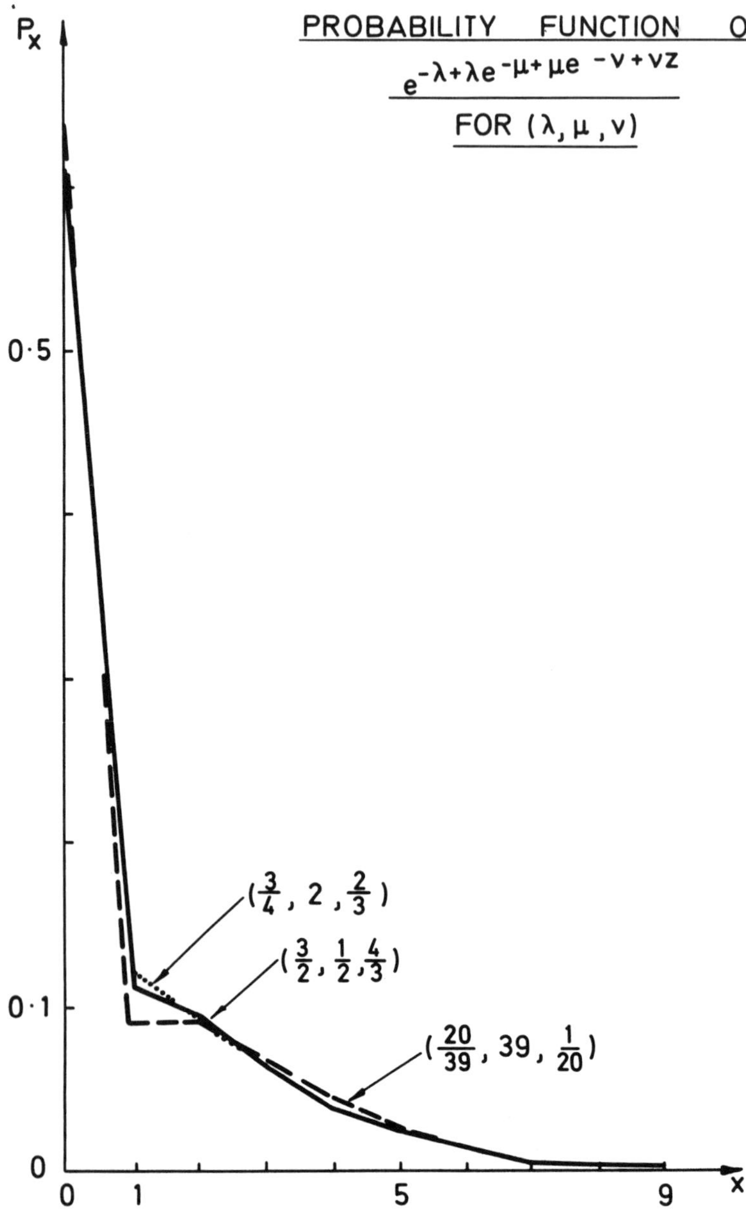
PROBABILITY FUNCTION OF
$e^{-\lambda+\lambda e^{-\mu+\mu e^{-\nu+\nu z}}}$
FOR (λ, μ, ν)
P_x
0·5
0·1
0
0
1
5
9
x
$(\frac{3}{4}, 2, \frac{2}{3})$
$(\frac{3}{2}, \frac{1}{2}, \frac{4}{3})$
$(\frac{20}{39}, 39, \frac{1}{20})$

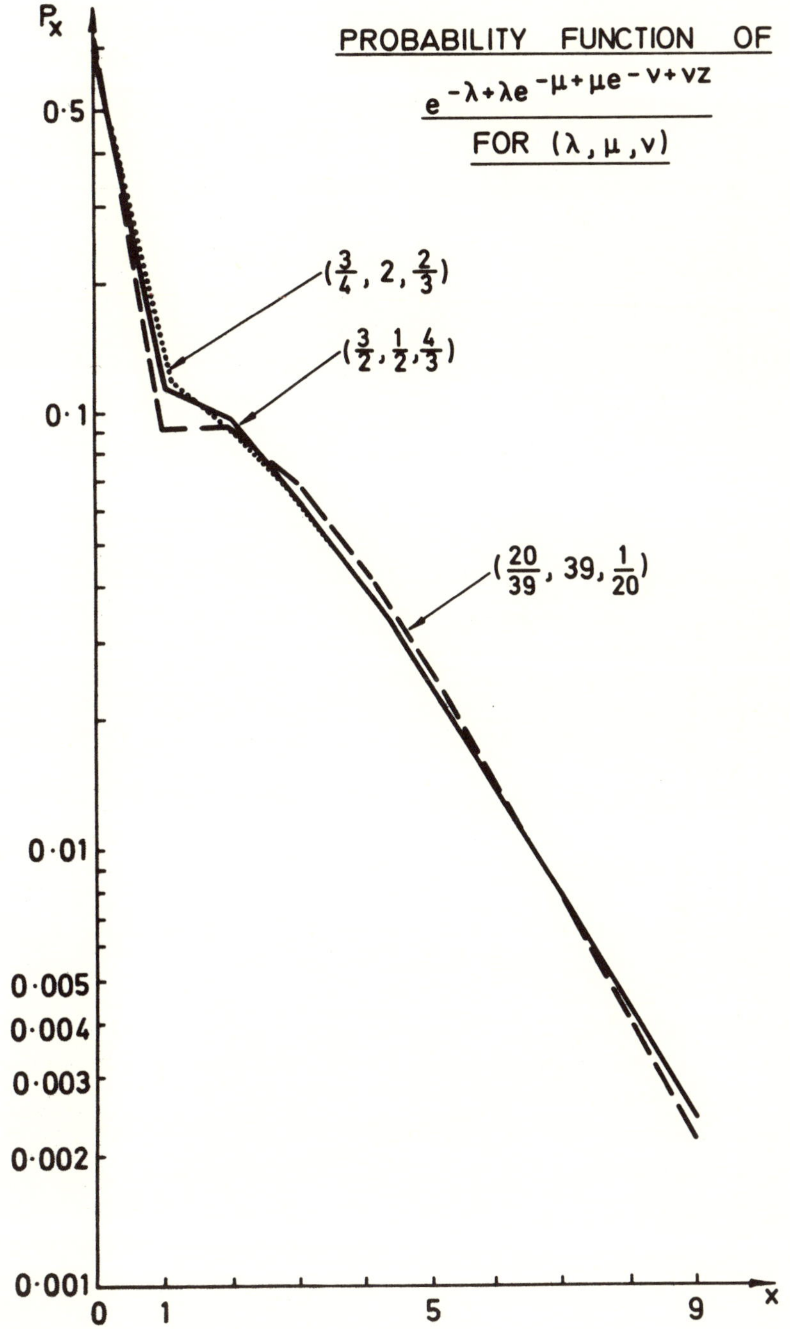
PROBABILITY FUNCTION OF
$\dfrac{e^{-\lambda+\lambda e^{-\mu+\mu e^{-\nu+\nu z}}}}{\text{FOR } (\lambda, \mu, \nu)}$
P_x
0·5
0·1
0·01
0·005
0·004
0·003
0·002
0·001
$(\frac{3}{4}, 2, \frac{2}{3})$
$(\frac{3}{2}, \frac{1}{2}, \frac{4}{3})$
$(\frac{20}{39}, 39, \frac{1}{20})$
0
1
5
9
x

$$\sum_{\nu} (\chi + \tilde{\omega} z)^{\nu} (1 - \alpha)\alpha^{\nu}$$

$$= \frac{1 - \beta}{1 - \beta z} \quad \text{for } \beta = \frac{\alpha\tilde{\omega}}{1-\alpha\chi}$$

$$(= g(\chi + \tilde{\omega} z)),$$

again a geometric distribution in which neither α nor $\tilde{\omega}$ is separately estimable.

A final subtlety of model construction is provided by the negative binomial distribution. (I owe this example to G. P. Patil.) Consider a sequence of successive equal intervals

...——'——————'——————'——————...

ith interval $(i + 1)$th interval

X_i events X_{i+1} events

during which independent numbers of events $X_1, X_2, \ldots$ occur, supposed logarithmic (α). In an experiment, the probability of N such intervals being observed is taken to be Poisson (μ). Then, as before, the total number of events

$$X = X_1 + X_2 + \ldots + X_N$$

is found to be described by

$$\left(\frac{1}{1-\alpha} - \frac{\alpha}{1-\alpha} z\right)^{\mu/\ln(1-\alpha)},$$

a negative binomial distribution. (If a logarithmic distribution with zeroes

$$\delta + (1 - \delta)\frac{\ln(1 - \alpha z)}{\ln(1 - \alpha)}$$

is used, the result is unchanged except that μ is replaced by $\mu(1 - \delta)$.) Suppose however that the interval scale is changed so that intervals of twice the length are observed. It is natural to take the number of events per interval to be Y_j, say, with a logarithmic distribution in which the mean is twice that of X_i, and the number P of intervals to be Poisson ($\frac{1}{2}\mu$). Of course this leads to a negative binomial distribution for

$$Y = Y_1 + Y_2 + \ldots + Y_P:$$

$$\left(\frac{1}{1-\beta} - \frac{\beta}{1-\beta} z\right)^{\mu/2 \ln(1 - \beta)},$$

where β is the parameter of the new logarithmic distribution and

$$\frac{-\beta}{(1-\beta)\ \ln\ (1-\beta)} = 2\ \frac{-\alpha}{(1-\alpha)\ \ln\ (1-\alpha)} \ .$$

But there is a contradiction here: if the ith and (i+1)th are two amalgamated intervals it is not the case that

$$X_i + X_{i+1}$$

has a logarithmic distribution: the convolution of two identical independent logarithmic variates has a Stirling distribution of the First Kind, not a logarithmic, contrary to the natural assumption about Y_j. However it is to be noticed that if X_i and X_{i+1} are supposed to have the logarithmic bivariate distribution

$$\frac{\ln[1 - \alpha(z_i + z_{i+1})]}{\ln\ (1 - 2\alpha)} ,$$

that of $X_i + X_{i+1}$ is

$$\frac{\ln(1 - 2\alpha z)}{\ln(1 - 2\alpha)} ,$$

again a logarithmic distribution. Thus it does not follow that the kind of application previously introduced, of specks from particles (or R. A. Fisher's [6] animals from species, or accidents per shift), encounters this difficulty: the "unit" of the interval is already defined, and no arbitrary division of it into independent units is possible.

No exact sampling distributions are usually available for any estimators in the compound distribution discussed here: but if attention is focused on comparisons of particular estimates as suggested to investigate the appropriateness of models, some standard errors or confidence intervals, even if of an approximate character, must be used. The precise algebraic form of these depends on the method of estimation (moments, maximum likelihood, minimum chi-square and various special methods), but if the usual large sample approximations following from the asymptotic approximations for an estimator T of some parameter with sample size n are

$$\mathcal{E}(T) = \alpha_o + \alpha_1/n + \alpha_2/n^2 + \ldots,$$

$$V(T) = \beta_1/n + \beta_2 n^2 + \ldots,$$

a rough 95% central confidence interval for the parameter is

$$(\hat{\alpha}_o \pm 2 \surd \frac{\hat{\beta}_1}{n}) ,$$

where $\hat{\alpha}_o$ and $\hat{\beta}_1$ are the functions obtained by replacing parameter values by the estimators. By reference to recent work (e.g., Shenton & Bowman [14]) it is sometimes possible to exercise a proper caution as to the sample size necessary for such an approximation to be used: roughly speaking, the terms α_1/n and β_2/n^2 should be small compared with α_o and β_1/n, respectively.

For example, if a random sample of size 100 comes from a Neyman Type A distribution—whatever the model!—and gives estimates of μ and ν of 1 and 5, respectively, reference to Shenton and Bowman [14] gives that if these are moment estimates the approximate 95% intervals are

for μ	for ν
(1 ± 0.4)	(5 ± 1.8)
$= (0.6, 1.4)$	$= (3.2, 6.8)$;

if maximum likelihood estimates

(1 ± 0.3)	(5 ± 0.9)
$= (0.7, 1.3)$	$= (4.1, 5.9)$.

The tables there given show that the higher order terms are not troublesome, except perhaps for a borderline case for the bias in the moment estimator for ν; the covariance between the estimators is ignored in writing down separate intervals for μ and ν. For estimators based on the components of the model (with which the above estimates might be compared) the method of estimation there would determine the details of such comparisons.

In modeling there is usually no difficulty in introducing new steps and new components, even if only to play safe by allowing for a possible kind of variation. No rules for dealing with the problem of when to stop are offered: as usual, one tries to introduce as few components (and parameters) as necessary, and what is 'necessary' and 'sufficient' for a descriptive purpose in one set of circumstances may be quite insufficient for analytical purposes in other circumstances. A balance has to be achieved between the mathematical complexities and adequacy of description, complicated by the variety of not implausible interpretations and problems of estimation—no general rules for this tentative scientific procedure are formulated, but lines along which the subject may be approached are suggested.

The arguments put forward here are all to the effect that in fitting complex distributions to random count frequency distributions, more attention ought to be given to the interpretation of the parameters (and less to goodness of fit). It is not argued that this is simple, nor indeed even that the model building process is always well defined or understood, but illustrations are given of modes of valid interpretation as well as of some difficulties (see also Evans [5] and Rogers [13]). (Although this is not intended to be a comprehensive review—for one reason why it is not, see the extensive literature review in Blischke [3]—one almost entirely omitted aspect might be referred to: the direct estimation of a mixing distribution, and the recent paper by Choi and Bulgren [4] gives a brief set of references.) If changes are made in the operating conditions, are these reflected in changes in the estimates as one would predict? If observations can be made on components directly, do these correspond with the overall estimates? If it should turn out that a certain class of distributions appears to describe well a wide variety of phenomena, from speck counts to cornborer larvae to insurance claims, this is economical; but it is hardly plausible to expect that merely from a frequency table of such counts one would make precise inferences about a model supposedly reflecting the structure in all of these cases. No more can be got from a model than has been put into in one form or another: if detailed observational material is not available on the structure of components it must be replaced by assumptions about the structure if assertions are to be made about these components.

This work was written during a visit to the Department of Statistics at The Pennsylvania State University while the author was on sabbatical leave from the University of New South Wales.

REFERENCES

[1] Anscombe, F. J. 1950. Sampling theory of the negative binomial and logarithmic series distributions. Biometrika 37:358-82.

[2] Arbous, A. G. and Kerrich, J. E. 1951. Accident statistics and the concept of accident-proneness. Biometrics 7:340-432.

[3] Blischke, W. R. 1965. Mixtures of discrete distributions. Classical and Contagious Discrete Distributions. (ed. G. P. Patil). Statist. Pub. Soc. (Calcutta) 351-72.

[4] Choi, K., and Bulgren, W. G. 1968. An estimation procedure for mixtures of distributions. J. Roy. Statist. Soc. (B). 30:444-60.

[5] Evans, D. A. 1953. Experimental evidence concerning contagious distributions in ecology. Biometrika 40:186-211.

[6] Fisher, R. A.; Corbet, A. S.; and Williams, C. B. 1943. The relation between the number of species and the number of individuals in a random sample from an animal population. J. Animal Ecol. 12:42-58.

[7] Jordan, C. 1965. Calculus of Finite Differences. Chelsea (N.Y.).

[8] Maceda, E. C. 1948. On the compound and generalized Poisson distributions. Ann. Math. Statist. 19:414-16.

[9] Molenaar, W. 1965. Some remarks on mixtures of distributions. Stichting Mathematisch Centrum, Amsterdam, Report S343.

[10] Newbold, E. M. 1927. Practical applications of the statistics of repeated events, particularly to industrial accidents. J. Roy. Statist. Soc. Ser. A. 90:487-

[11] Pielou, E. C. 1957. The effect of quadrat size on the estimation of the parameters of Neyman's and Thomas' distributions. J. Ecol. 45:31-47.

[12] Rao, C. R.; Mitra, S. K.; and Matthai, A. 1966. Formulae and Tables for Statistical Work. Statist. Pub. Soc. (Calcutta).

[13] Rogers, A. 1969. Quadrat Analyses of Urban Dispersion. Working paper no. 93, Center for Planning and Development Research, U. C. Berkeley.

[14] Shenton, L. R., and Bowman, K. O. 1967. Remarks on large sample estimaters for some discrete distributions. Technometrics 9:587-98.